PLANNING SUSTAINABLE AND RESILIENT FOOD SYSTEMS

Covid-19 was a canary in a mine. It exposed the vulnerabilities of 21st-century food systems but did not create them. Since then, the world has faced a "polycrisis:" a cluster of weather-related crop failures, war-induced food and energy shortages, and import dilemmas with compounding effects. Going forward, we need to plan for more sustainable and resilient food systems that improve environmental outcomes and address economic disparities. But food systems planning is a relatively new discipline and guidance is scarce. This book fills that gap.

Where most food systems planning has focused on urban issues, this book takes a holistic view to include rural communities and production agriculture whose stewardship of the earth is so critical to public and environmental health, as well as to ensuring a varied and abundant food supply. Its goal is to inform planning practices and follow-up actions for a wide range of audiences—from professional planners, planning commissions, and boards to conservation districts and Cooperative Extension to the on-the-ground change-makers working to strengthen America's food and farming systems. Embracing the fact that the U.S. is highly diverse in its people, places, and politics, the book lifts up principles and successful examples to help communities develop strategies based on their unique assets and the needs and preferences of their people.

Julia Freedgood is Senior Fellow and Senior Program Advisor at American Farmland Trust, where she focuses on planning, policy, and programs to support farmers and ranchers, protect farmland for farming, and advance sustainable and resilient food systems.

PLANNING SUSTAINABLE AND RESILIENT FOOD SYSTEMS

From Soil to Soil

Julia Freedgood

Routledge
Taylor & Francis Group
NEW YORK AND LONDON

Designed cover image: Shawn Linehan, Jacob Tosch, David Kosling, Preston Keres, and USDA.

First published 2025
by Routledge
605 Third Avenue, New York, NY 10158

and by Routledge
4 Park Square, Milton Park, Abingdon, Oxon, OX14 4RN

Routledge is an imprint of the Taylor & Francis Group, an informa business

© 2025 American Farmland Trust

The right of American Farmland Trust to be identified as author of this work has been asserted in accordance with sections 77 and 78 of the Copyright, Designs and Patents Act 1988.

All rights reserved. No part of this book may be reprinted or reproduced or utilised in any form or by any electronic, mechanical, or other means, now known or hereafter invented, including photocopying and recording, or in any information storage or retrieval system, without permission in writing from the publishers.

Trademark notice: Product or corporate names may be trademarks or registered trademarks, and are used only for identification and explanation without intent to infringe.

Library of Congress Cataloging-in-Publication Data
Names: Freedgood, Julia, author.
Title: Planning sustainable and resilient food systems : from soil to soil / Julia Freedgood
Description: New York, NY : Routledge, 2024 | Includes bibliographical references and index
Identifiers: LCCN 2023056945 (print) | LCCN 2023056946 (ebook) | ISBN 9781032276854 (hardback) | ISBN 9781032276861 (paperback) | ISBN 9781003293712 (ebook)
Subjects: LCSH: Agriculture and state—United States | Food supply—United States | Food security—United States | Sustainable agriculture—United States
Classification: LCC HD9006 .F74 2024 (print) | LCC HD9006 (ebook) | DDC 338.1/973—dc23/eng/20240405
LC record available at https://lccn.loc.gov/2023056945
LC ebook record available at https://lccn.loc.gov/2023056946

ISBN: 9781032276854 (hbk)
ISBN: 9781032276861 (pbk)
ISBN: 9781003293712 (ebk)

DOI: 10.4324/9781003293712

Typeset in Joanna
by Apex CoVantage, LLC

*To Peggy Rockefeller, Norm Berg, and Ralph Grossi,
who individually and together laid the foundation
and inspired this work.*

*And to the unsung heroes: the farmers, ranchers, and farm
workers who produce our food and the essential workers who
help get it from field to fork.*

CONTENTS

About the Author   ix
Foreword   xi
Acknowledgments   xiii
About American Farmland Trust   xvi

Section I

Introduction   3

1   Why Plan for Food and Agriculture?   25

2   The Public Framework for Food Systems Planning   51

3   Principles and Practices to Guide Planning for Food and Agriculture   77

Section II

4   Federal Policies That Affect Food Systems   105

5   Land Use Policies That Support Farms, Farmland, and Food   121

6	**Programs and Policies to Sustain Agriculture**	157
7	**Programs and Policies to Support Community Food Security**	192
8	**Evolving Issues**	217
	Appendix: Organizational Resources	243
	Bibliography	247
	Index	264

ABOUT THE AUTHOR

Julia Freedgood is Senior Fellow and Senior Program Advisor at American Farmland Trust, where she focuses on planning, policy, and programs to support farmers and ranchers, retain and protect farmland for farming, and advance sustainable and resilient food systems.

Freedgood is a national expert in land use and protection, land tenure and access, agricultural viability, and food system planning. Since 1986, when she helped launch the seminal Massachusetts farmers market coupon program, she has devoted her career to finding practical solutions to complex problems affecting local and regional food systems, conducting research, providing technical assistance, and actively engaging in community planning and capacity building.

She is the author or co-author of many articles and reports; recent work includes "Farms Under Threat: The State of the States," "Fail to Include, Plan to Exclude: Reflections on Local Governments' Readiness for Building Equitable Community Food Systems," and "Growing Local: A Community Guide to Planning for Agriculture and Food Systems."

A longtime member of the American Planning Association, Freedgood also is a founding member of North American Food Systems Network and serves on the Advisory Board and Editorial Committee of the *Journal of*

Agriculture Food Systems and Community Development. She has served on numerous USDA advisory boards and grant review committees, including USDA-Secretary Vilsack's Land Tenure Advisory Committee. She holds a master's degree in urban and environmental planning and policy from Tufts University.

FOREWORD

Thirty years ago, in the small Maine town where I lived, I found myself co-chairing the comprehensive plan committee with a local dairy farmer. The State of Maine at the time required every community to develop such a plan—and a group of us were intent on crafting a forward-looking vision that would enhance our town's rural character and help retain farms. At a time when few people saw a future for farming in northern New England, it took dozens of public meetings over several years to educate and build consensus. The work was arduous. We often felt that with every step forward, we took two steps back. But the committee soldiered on, and in the end, the plan that emerged—which put farming front and center—led the community to enact a new land use ordinance that made a difference locally.

This experience in community planning opened my eyes to farming's potential as both a local economic engine and a powerful tool to achieve essential environmental goals, from maintaining open space to protecting wildlife and water supply. It placed me squarely on a professional path advancing agriculture. I initially focused on helping Maine farmers improve business operations, often by entering new markets. This work inevitably led me to realize that farm viability, as important as it is, could not alone counter mounting development pressures. Motivated by a report on farmland loss from American Farmland Trust, a group of us created Maine Farmland Trust. Like many other nonprofits organized to save farmland and support land access, Maine Farmland Trust did and continues to do critically important work. Yet these strategies by themselves are not enough.

Since its founding in 1980, American Farmland Trust has stressed how agriculture is a system—a system intertwined in the most fundamental way with how we sustain ourselves and our planet. For over three decades, Julia Freedgood has been a stalwart and impassioned advocate of this holistic view. In this volume, she explains how agriculture is not only essential to a sustainable food system, but to combating climate change. She describes the tools needed to incorporate agriculture into state and local plans that all too often ignore it. She helps us comprehend why we must be thinking of agriculture in all that we do, because, quite simply, our future depends on it.

John Piotti
President & CEO
American Farmland Trust

ACKNOWLEDGMENTS

To Peggy Rockefeller, Norm Berg, and Ralph Grossi, who individually and together laid the foundation and inspired this work. And to the unsung heroes: the farmers, ranchers, and farm workers who produce our food and to the essential workers who help get it from field to fork.

Special thanks to American Farmland Trust President John Piotti for visionary leadership, believing in me, and investing in this idea. Humble thanks to Otto Doering, who bravely reviewed my first draft and offered thoughtful, much-needed feedback. And heartfelt appreciation to Jennifer Dempsey and Cris Coffin, who contributed ideas, insights, inspiration, and lots of information along the way. This book would not have happened without them.

Thanks also to Farmland Information Center staff: Don Buckloh, Kayla Donovan, Kate Rossiter Pontius, and Ariel Looser for research, review, fact finding, fact checking, and so on; to American Farmland Trust colleagues Jim Hafner, Jerry Cosgrove, Samantha Levy, Caro Roszell, Bianca Moebius-Clune, and Billy Van Pelt for valued input, review, and feedback; Olivia Fuller and Kirsten Ferguson for help finding photos; Ryan Murphy for providing *Farms Under Threat* maps; and Sam Smidt for the rural employment graph.

Profound gratitude to my partners in two collaborative projects—*Growing Food Connections* and *Pathways to Prosperity*—who mixed scholarship with practical wisdom to reimagine food systems and how to approach planning for them.

The book is infused with years spent percolating, steeping, and ruminating with them through monthly Zoom meetings, annual get togethers, research, and writing. First to Jill Clark whose game-changing work centers on governance, the policy process, and community engagement. To Samina Raja and Kim Hodgson for their acumen, aptitude, and illuminating the urban perspective, and to Jeanne Leccese for her grounded perspective and for moving GFC forward. Likewise to Shoshanna Inwood and Becca Jablonski, for inspired scholarship and the fire in their bellies when they contemplate what it means to be rural. Finally, to Aiden Irish for his research and masterful management of P2P.

> *Growing Food Connections* was a 5-year research, extension, and education project to enhance community food security while ensuring sustainable and economically viable agriculture and food production. It focused on building local government capacity to remove public policy barriers and deploy innovative policy tools, and was supported by a USDA National Institute of Food and Agriculture Food Systems Program award (grant no. 2012-68004-19894). Learn more at https://growingfoodconnections.org/.
>
> *Pathways to Prosperity* was a 4-year applied research and extension project focused on collaborative approaches to strengthening the value-added food and agricultural sector in rural communities. It was supported by Innovation for Rural Entrepreneurs and Communities Program (grant no. 2019-68006-29681) from the USDA National Institute of Food and Agriculture. Learn more at https://localfoodeconomics.com/pathways/.

Sincere thanks for the data, maps, photographs, publications, and everything I learned through emails and interviews with the following dedicated and knowledgeable people:

- Nyssa Entrekin, Caroline Haynes, and Julia Koprak from The Food Trust;
- Bill Hoffman and Ashley Mueller from USDA National Institute of Food and Agriculture;
- Doug Jackson from Ohio State University;
- Jimmy Kroon and Eric Reid from the Delaware Agricultural Lands Preservation Program;
- Chris Mathias from the Scott County, Iowa, planning department;

- Beth Cawood Overman and Dustin Baker from Lexington-Fayette Urban Council of Governments;
- Susan Payne and Steven Bruder from the New Jersey State Agriculture Development Committee; and
- Susan Whitfield from No More Empty Pots.

Last but not least, a huge thank you to Sally Murray James of Cutting Edge Design for her fabulous work designing and managing graphic images for the book.

ABOUT AMERICAN FARMLAND TRUST

American Farmland Trust (AFT) works to save the land that sustains us by protecting farmland, promoting environmentally sound farming practices, and keeping farmers on the land. AFT is the only national agricultural organization that approaches its work in this comprehensive, holistic manner, combining on-the-ground projects with objective research, extension, and advocacy. We launched the conservation agriculture movement and continue to raise public awareness through our *No Farms No Food* message. Since our founding in 1980, AFT has helped permanently protect about eight million acres of agricultural land, advanced environmentally sound farming practices on millions more acres, and supported a half million farm families. Learn more at www.farmland.org.

Section I

INTRODUCTION

Covid-19 was a canary in a mine. It exposed the vulnerabilities of 21st-century food systems but did not create them. When the pandemic hit, American consumers were shocked to find empty shelves at supermarkets—appalled when packing plant closures caused meat shortages, leaving livestock producers without processing alternatives, and still at least 59,000 meatpacking workers contracted the virus in the pandemic's first year. The outbreaks disproportionately affected low-wage workers from racial and ethnic minorities and led to community spread in rural areas.[2] Meanwhile, rising food costs combined with high unemployment increased hunger, especially among children. While failing to protect health and safety or to fairly compensate their workers, the top meatpacking companies earned record profits.[3]

These are examples of some of the unintended consequences of decades of U.S. food and farm policies. Driven by the "get big or get out" philosophy famously expressed by 1970s Agriculture Secretary Earl Butz, these policies encouraged specialization and consolidation across food and agricultural sectors. In many ways, they were very successful. They achieved their goal of increasing production and distribution efficiencies to produce a plentiful and cheap food supply, successfully expanding global markets and American's food choices. Total farm output nearly tripled between 1948 and 2019[4] and food prices fell. Today, on average, most consumers only spend about 10 percent of their disposable personal income on food[5] (see Figure 0.1). However, cheap food does not ensure public or environmental health.

In 2021, nearly 34 million Americans lived in food-insecure households[6] and 61 percent of Supplemental Nutrition Assistance Program (SNAP) participants could not afford to buy healthy food.[7] Driven by economic disparities and barriers to food access, food insecurity is most prevalent in rural areas and large cities and its impacts felt most severely by Black, Hispanic, and single-mother households.[8] Rural food insecurity has had an outsized impact on Indigenous communities, especially during Covid. One study found that 56 percent of American Indian/Alaskan Natives (AIAN) experienced food insecurity, and nearly a third reported very low food security.[9] Further, cheap food is not necessarily nutritious. A Harvard School of Public Health meta-analysis found healthier diets cost about $1.50 more

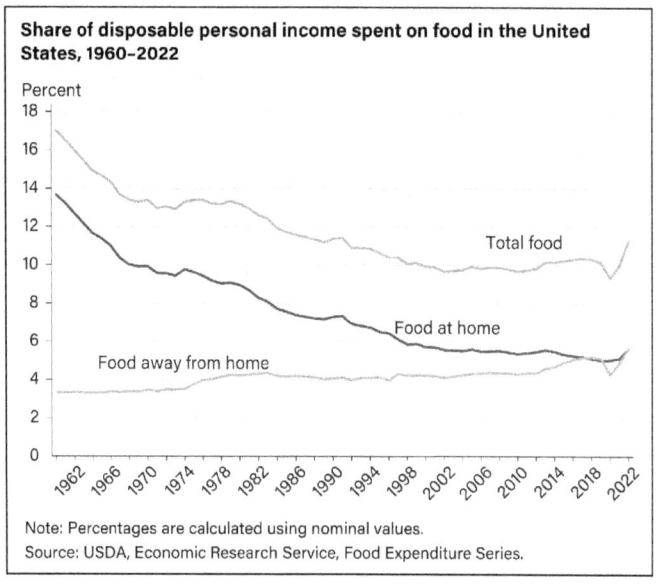

Figure 0.1 Consumer spending on food fell sharply in the second half of the 20th century.

per day than unhealthier, likely due to policies focused on producing high-volume commodities and a supply chain that favors increasing profitability for highly processed food products.[10]

Farm production efficiencies emphasized specialization. This separated animal and crop production, increasing the need for synthetic fertilizers and pesticides, antibiotics to boost livestock growth, and large lagoons for livestock manure. These inputs add costs to agricultural operations, degrade the environment, and can harm human health.[11] It also drove increased crop production for feed and fuels, especially corn.

Today the U.S. is the world's largest producer, consumer, and exporter of corn. U.S. farmers plant an average of about 90 million acres each year, mostly for livestock feed and ethanol. Some also is processed for human consumption for cereals, flour, grits, and beverage alcohol, and products like high fructose corn syrup, glucose, dextrose, corn starch, and oil[12] (see Figure 0.2).

Corn—or maize—has been cultivated in the Americas for thousands of years. A versatile and valuable crop, its dramatic expansion came with significant environmental costs,[13] which often are used to symbolize agriculture's negative impact on the environment. Corn is not alone. Since 1962, when Rachel Carson published *Silent Spring*, the environmental impacts of modern

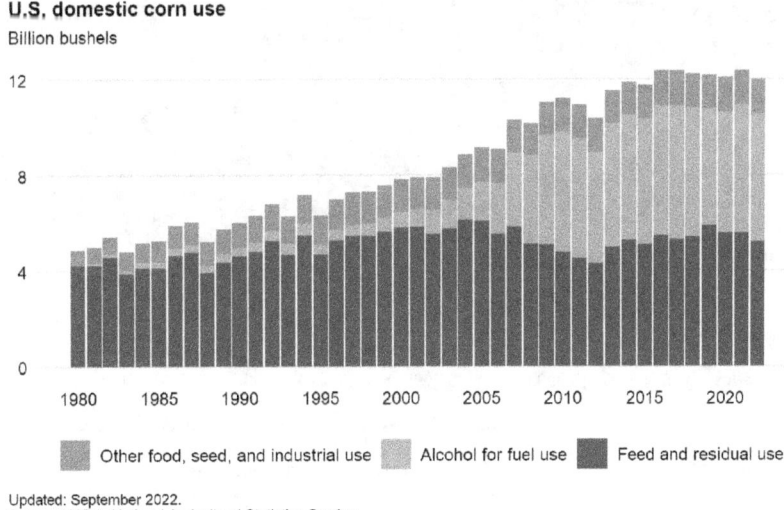

Figure 0.2 Domestic corn increasingly is grown for ethanol along with feed grain use.

agriculture have been widely documented: degraded soil and water quality, elevated greenhouse gas (GHG) emissions, and loss of biodiversity.[14]

Farm consolidation accompanied specialization. Where once farmers were diversified and grew multiple crops, today most grow two or three. Poultry and livestock operations increasingly rely on purchased feed instead of growing their own. Cropland ownership and agricultural wealth are ever more concentrated on a smaller number of highly specialized farms in rural areas, where less than a fourth of the population lives.[15] In 2015 more than two-thirds of production came from only 5 percent of farms that had at least $1 million of output[16] (see Figure 0.3).

The profound shifts of the 20th century came at the expense of mid-sized farms, rural communities, and diversified supply chains. It does not have to be this way. But it will take a planning and policy shift as transformational as "get big or get out" to promote food systems that are at once sustainable, resilient, and fair.

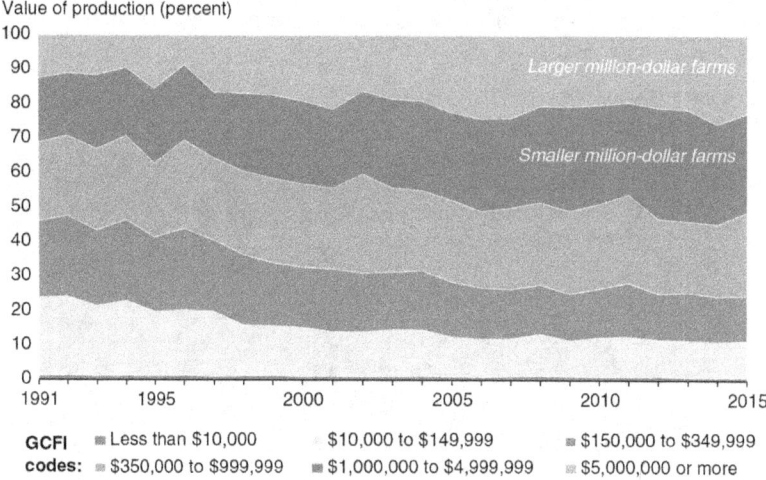

Figure 0.3 Farm production shifted to farms with annual sales of $1 million or more.

Small and midsized farms typically employ more people, create more cohesive communities, and contribute more to local economies than very large farms.[17] They are more likely to be diversified and to purchase livestock, equipment, and supplies locally.[18] Over the 20th century, the total number of farms fell more than 60 percent, while average farm size increased by 67 percent,[19] hollowing out rural communities and making it harder for small commercial and midsized farms to compete. The trend continues. In the 2017 Census of Agriculture, all farm numbers declined except for the very smallest and very largest.[20] Between 2017 and the 2022 Census, small- and medium-sized farms declined while large farms with sales of $1 million or more increased. In 2022, large farms only represent six percent of all farms, they sold more than three fourths of all agricultural products.[21]

These losses disproportionately harmed racial minorities. In 1920, nearly 15 percent of farmers were Black, Indigenous, or Asian.[22] In the 2022 Census of Agriculture, the number of farmers of color is only 4.6 percent[23]—a striking contrast to the general population, which is increasingly diverse and where 42 percent identifies as a race or ethnicity other than "White alone."[24] In particular, Black, Hispanic, and Indigenous farmers were pushed out of agriculture by redlining, heirs and fractionated property rights disputes,[25] direct violence, and discriminatory USDA lending policies, which ultimately led to billions of dollars in payouts to settle class-action lawsuits.

Consolidation has also occurred throughout the supply chain. Food industries are increasingly concentrated. A handful of transnational companies dominate domestic food supplies, with four or fewer firms controlling at least 50 percent of the market for 79 percent of common groceries.[26] In 2021, the top four food retailers commanded more than half of all sales[27]—more than doubling their share since 1997.[28] This also is true for processing, especially of livestock. In the beef market, four companies control 85 percent of purchasing and processing; in pork, they control 67 percent.[29]

It Is Time to Shift the Paradigm

The stress the pandemic placed on food supplies exposed the brittleness of "get big or get out." It illuminated the importance of the people who provide our food—from the farmers, ranchers, and farm workers who grow it to the people throughout the supply chain who process, distribute, and serve it. It revealed inherent inequities, including that nearly half of the 25 lowest-paying jobs in the U.S. are related to food systems.[30] And it pointed to the need for comprehensive food systems planning.

Most people talk about THE food system as if it were a single thing. In truth, we participate in a myriad of intersecting food systems,[31] nested together like Matryoshka dolls, operating at many levels from emergency food to global supply chains, with community, local, regional, tribal, and domestic food systems nestled in between. Each is important; each has a role to play. Food systems are multidimensional, social-ecological systems. They rely on natural resources like arable land, clean water, sunshine, and various microclimates to produce diverse crops and livestock. They are sustained by the management and labor of farmers, ranchers, farm workers, and a vast assortment of intermediaries who handle food between field and fork. And they are influenced by many things, including dietary preferences, human behavior, public policies, and market forces like supply and demand—the

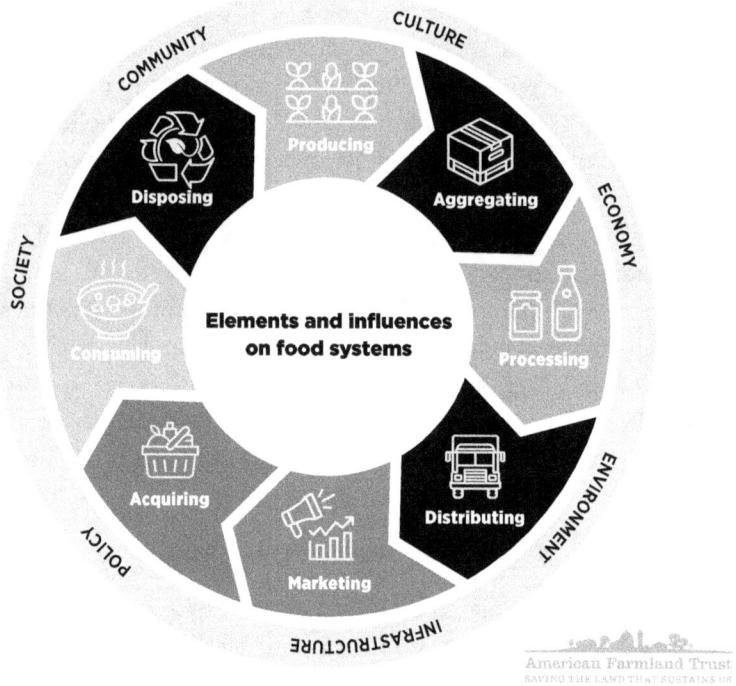

Figure 0.4 Food systems encompass the life cycle of food from cultivation through consumption to the disposal of waste.

dynamic push and pull between producers, supply chain firms, and consumers. Impacts affecting one part of the system affect the rest, along with the price of food.[32]

U.S. food and farm policy largely is driven by the federal government with food supplies left to the private market. Recently, state and local governments have become more engaged, especially in food systems planning. A relatively new discipline, it emerged at the turn of the century after two planning professors, Jerome Kaufman and Kami Pothukuchi, wrote a couple of seminal articles urging planners to pay more attention to food.[33] Since then, interest has grown so much that the American Planning Association (APA) formed a Food Systems Division (FSD) and Michigan State University researchers found that in 2021 nearly 60 percent of states have created some kind of food plan or had one under development.[34]

Most food systems planning efforts focus on food access, nutrition, and urban agriculture. Driven by mounting evidence that the earth has reached a tipping point, it is time to take a more holistic approach to address food security in the context of smart growth, regenerative agriculture, and community resilience. Beyond the pandemic, speculative investment in commodity markets, global political unrest, and the impacts of climate change are disrupting agricultural production and food supplies at home and abroad.

The U.S. is home to over 10 percent of the planet's arable land.[35] Yet even here, in what appears to be a vast agricultural landscape, less than 20 percent of the continental U.S. comprises nationally significant soils—the farmland best suited for sustainable food production[36] (see Figure 0.5).

Poorly planned development threatens the best farmland in every state. From 2001 to 2016, the U.S. converted 11 million acres of agricultural land—equivalent to all the land planted to fruits, nuts, and vegetables in 2017.[37] Beyond its importance for food production, agricultural land produces fewer GHG emissions than land converted to housing or commercial uses.[38] And when managed using regenerative or climate-smart practices, it serves as a natural carbon sink by drawing down CO_2 and storing it in plants and soils. At a time when growing demands for healthy food are colliding with the environmental impacts of a changing climate, the food systems that depend on this vital and irreplaceable resource are threatened by competition for land, market consolidation, and an aging farming population. It is time for a more holistic approach to food systems planning.

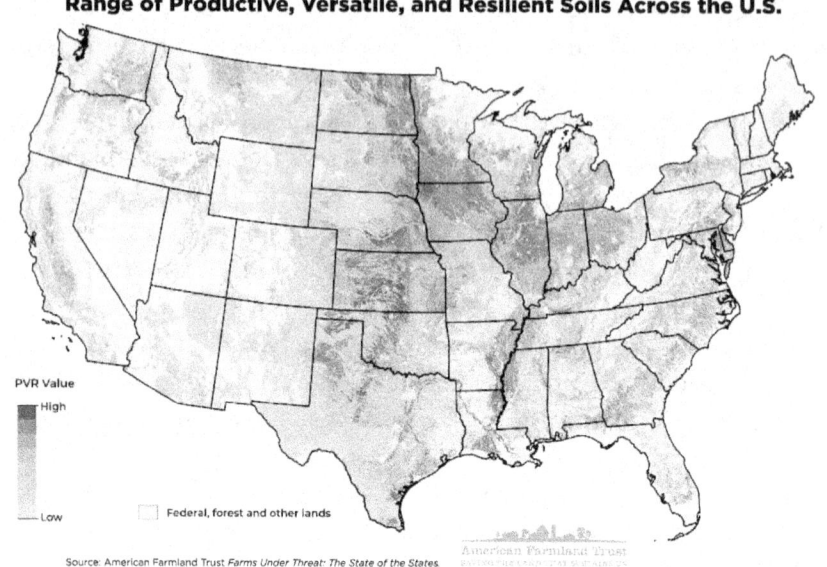

Figure 0.5 This map identifies nonfederal land best suited for long-term cultivation, especially for food crops like fruits, nuts, vegetables, and staple grains.
Source: American Farmland Trust, "Farms Under Threat: The State of America's Farmland."

Since 1988, the Intergovernmental Panel on Climate Change (IPCC) has documented impacts on food security from rising temperatures, increased frequency of extreme weather events, and other impacts of a changing climate. It projects that without intervention, these patterns will intensify, with especially significant impacts on low-wealth and vulnerable communities, increasing food insecurity in a world where the population is expected to reach nearly 10 billion by 2050.[39] And it points to agricultural practices that can be scaled up to increase resiliency, including regenerative agriculture, diversification, and integrated production systems.[40]

Extreme weather events are increasing in frequency and intensity,[41] and the eight years between 2015 and 2022 were the warmest ever recorded.[42] In some cases and places, warmer temperatures may increase crop yields, but overall the changing climate is projected to reduce food supplies, raise food prices,[43] and lower important nutrients in crops.[44] Fruit and vegetable crops are especially sensitive to climatic stressors as heat waves and cold snaps lead to lower yields and/or reduced quality, often with devastating impacts on profitability. Water shortages and deteriorating water quality pose additional threats.[45]

Greenhouse gas (GHG) emissions, which trap the sun's heat in the atmosphere, are driving these dramatic changes. In the U.S., their main source is burning fossil fuels for electricity, heat, and transportation. But food systems activities also generate GHG emissions from producing, processing, refrigerating, transporting, and disposing of food in landfills. Agriculture contributes about 10 percent of total U.S. emissions,[46] largely due to methane generated from liquid manure in confined animal feeding operations (CAFOs) and certain kinds of cropping systems[47] (see Figure 0.6).

For millennia, people have used plows to till or cultivate the soil. Most crop production still uses tillage to break up and turn over soil to plant seeds and control weeds. But tilling kills beneficial microorganisms that build and sustain the soil. We need healthy soils to produce a plentiful and nutritious food supply. We also need them to support ecosystem services, from erosion

Sources of U.S. Greenhouse Gas Emissions in 2021

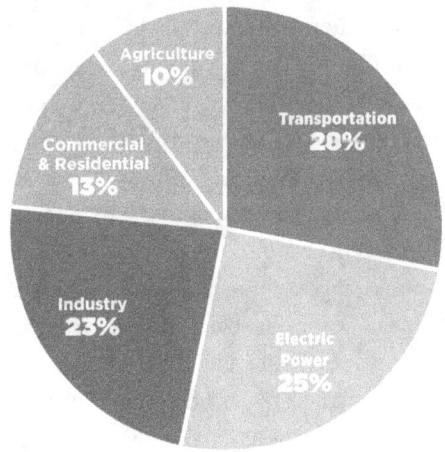

Source: U.S. Environmental Protection Agency emission estimates from the Inventory of U.S. Greenhouse Gas Emissions and Sinks: 1990–2021.

Figure 0.6 The largest source of U.S. greenhouse gas emissions from human activities comes from burning fossil fuels for electricity, heat, and transportation. Agriculture contributes about 10 percent, mostly from livestock, agricultural soils, and rice production.

and flood control to filtering and buffering pollutants, to pollinator and wildlife habitats.

Farming practices that build soil health are vital for continued productivity and to sequester carbon, thus serving as an immediate, cost-effective approach to combating climate change. Some farmers already are adopting these practices. According to USDA Natural Resources Conservation Service (NRCS), voluntary conservation efforts have increased carbon stored in cultivated cropland soils by nearly 9 million tons annually.[48] If all the world's farmers and ranchers used these practices, it is estimated they could sequester enough carbon annually to offset up to 20 percent of global fossil fuel emissions.[49] Going forward, communities as well as countries must work harder to prepare for, manage, and recover from food systems stresses, disruptions, and shocks. Drawing on the 2022 IPCC report, we all must improve our capacity for adaptation, learning, and—perhaps most importantly—transformation.[50]

This book is an attempt to do just that. Its goal is to inform planning practices and follow-up actions for a wide range of audiences: professional planners, planning commission and board members, elected officials, Extension educators, conservation district staff, and all the on-the-ground changemakers who want to strengthen America's food and farming systems. It celebrates planning to lift up the big picture and bring people together to meet community goals and offers guidance on steps we can take and tools we can use to transform food systems to improve economic, environmental, and social outcomes.

The book focuses on land-based systems, starting with agriculture's role in ensuring the U.S. can deliver an affordable and nutritious food supply while improving environmental outcomes and addressing economic disparities. It appreciates—but does not emphasize—the importance of foraging, hunting, gathering, and marine and freshwater sources to food security and cultural identity. These also are part of the food systems mix. And it recognizes that systems solutions are multidimensional, cross-disciplinary, and complex.

Transforming food systems cannot be solved sector by sector. Yet most U.S. food and farm policy is piecemeal, driven by the omnibus federal Farm Bill, which addresses the different parts of the system largely without connecting the whole. Generally reauthorized every five years, as of this writing, Congress has yet to authorize the 2023 Farm Bill, but it is not expected to be transformational. The 2018 bill was extended through September 2024 and includes 12 titles. Nutrition receives about three-fourths of direct funding, followed by agricultural commodities, crop insurance, and conservation. Less than .01 percent of its total spending goes toward programs that connect nutrition and farm production.[51] Given its history and the powerful interests it serves, federal policy will be hard and slow to change.

State and local governments can address food systems on their own and by tapping federal program support. Food systems planning is a good place to start. But it must include production agriculture whose stewardship of the earth is so critical to public and environmental health, as well as to domestic food supplies. This will require incorporating rural planning and agriculture into food systems strategies along with urban.

Planning with and for rural communities is vital to ensuring nutritional security at multiple levels. Most have arable land and working farms and ranches that produce most of our staple food crops and livestock. In these places, agriculture is the bedrock of rural economies, representing significant employment and having a far greater impact per employee on the local economy than in urban areas (see Figure 0.7). However, the planning profession was developed largely to guide land use and development in cities.

Most cities have dedicated planning departments and resources that focus on the built environment. When they address food systems, they tend to rely on zoning and local land use regulations instead of investing in the middle infrastructure needed to connect production with consumption and build larger, more diversified markets.[52] On the other hand, rural communities often lack resources for any kind of planning, much less food systems planning. They rarely address food production, food access, or nutritional security. Yet the counties with the highest rates of food insecurity are disproportionately rural.[53]

One of the challenges with rural planning is knowing how to define it. Federal agencies use a dizzying array of more than two dozen definitions. They have one thing in common: They define rural by what it lacks rather than what it is. The U.S. Census consigns rural to people, housing, and territory outside of an urban area. Even USDA definitions rely on an absence of urban features—like population or adjacency to populated areas—instead of the presence of rural assets.[54] Given low population density, rich natural resources, but limited built infrastructure, rural needs and opportunities often are different from urban. But they are just as important to food systems planning.

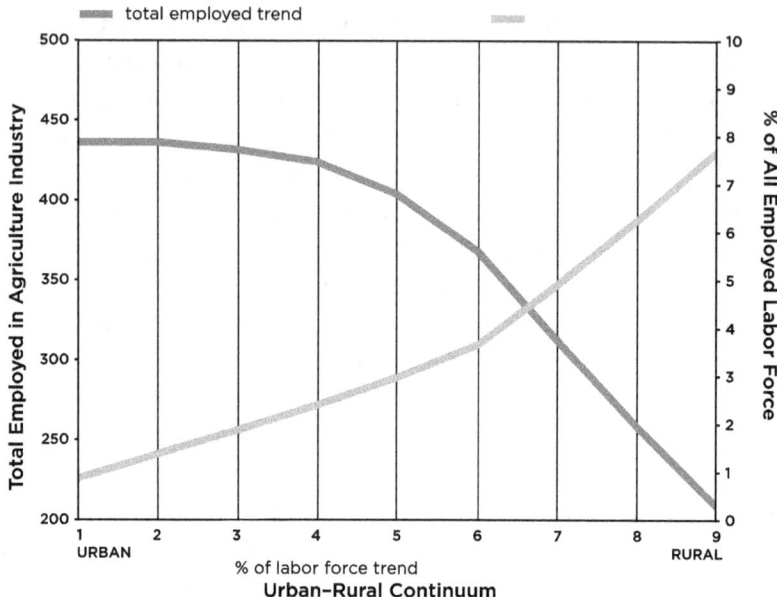

Agricultural employment plays a proportionately higher role in rural than urban counties.

Source: American Farmland Trust Land Use Policy Research using USDA Economic Research Service 2021 Atlas of Rural and Small-Town America and 2013 Urban Influence Codes

Figure 0.7 Agricultural employment in rural areas has fewer total jobs than in urban areas but reflects a greater percentage of all employed and has a greater impact per employee on the local economy.
Source: American Farmland Trust Land Use Policy Research division.

USDA-ERS's Rural Urban Continuum (RUC) Codes provide a useful framework for understanding these relationships. They classify urban—or metropolitan—counties by the population of their entire metro areas, and nonmetropolitan counties by degree of urbanization and adjacency to a metro area. Many "metro" counties have vibrant agricultural and local food economies where most of our local and perishable foods are grown. They still have rural features and are a vital link in the food supply chain, supplying two-thirds of direct-to-consumer (DTC) food sales, mostly within a 100-mile radius of the farm.[55]

Direct Sales of Food Produced in Urban-Influenced Counties
By Percent of Market Value

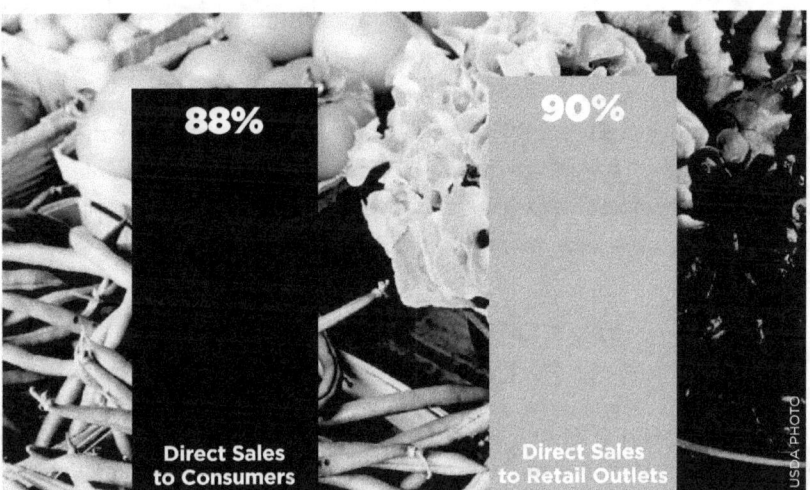

Figure 0.8 Most local and perishable food is grown in urban-influenced counties. Source: American Farmland Trust Farmland Information Center.

Going forward, agriculture and food production will and must occur in all kinds of environments: rural, urban, and along the increasingly urban-influenced continuum in between. Thus, planning for food systems requires a holistic approach to create regional linkages and support for more diversified markets and infrastructure. This will lead to healthy symbiotic relationships along the continuum and across the food supply chain. But there are very few resources to guide this process.

This book seeks to fill that gap. It presents pathways to achieve sustainable, resilient, and evenhanded food systems regardless of scale or scope. Written with the hope that with better understanding and by working together we can bridge the growing urban–rural divide, the book embraces the fact that the U.S. is highly diverse: In its people, its places, its politics—even in the powers invested through states to local governments. Since what is best in one community may be detrimental to another, it lifts up principles and successful examples, but eschews "best practices." Strategies and solutions must be specific to their context, connected to each community's natural and cultural resources, and to the needs and preferences of its people.

Finally, while policy change is an important lever, transformation is a collective process. Planning lays the groundwork; it sets it in motion. To be effective, food systems planning must be as inclusive as possible: engaging multiple perspectives, forms of knowledge, and cross-sector approaches. Since this requires clear communication, including a shared use of terms, let's start there.

Common Terms

People and disciplines often use the same words to mean different things—or different words to mean the same things, like hoagies, subs, and grinders. Sometimes words widely used in one discipline are adopted by another. For example, a common term in psychology, resilience increasingly is associated with climate science. They are equally important. While words can bring people together, they also can pull them apart, obscuring meaning and opportunities for common ground. To offer a common denominator, what follows are the meanings used for some key terms used frequently in this book.

Agriculture

Agriculture is a set of activities used to produce crops, poultry, and livestock. Commonly associated with food, feed, fiber, flowers, and fuel, it also produces nursery plants and trees; ingredients for pharmaceuticals, cosmetics, and industrial alcohol; and other products. Most laws confine it to commercial production, and some states and policies expand the definition to include timber, raising fish (aquaculture), on-farm education and recreation (agritourism), growing plants in water using mineral nutrients (hydroponics), horse breeding, and even riding stables (equine). Beyond cultivation, it includes marketing and activities that add value to farm products, like processing fruit into jam.[56]

Organic, Sustainable, and Regenerative Agriculture

Organic, sustainable, and regenerative agriculture have overlapping goals and approaches but are not synonymous. According to USDA, *organic agriculture* involves

> the application of a set of cultural, biological, and mechanical practices that support the cycling of on-farm resources, promote ecological balance, and conserve biodiversity. These include maintaining or enhancing soil and water quality; conserving wetlands, woodlands, and wildlife; and avoiding use of synthetic fertilizers, sewage sludge, irradiation, and genetic engineering.[57]

Sustainable agriculture is an integrated system of plant and animal production practices with site-specific applications to achieve long-term social, economic, and environmental benefits. These include meeting human food and fiber needs, enhancing environmental quality, sustaining farm viability, and improving the quality of life for farmers and society as a whole.[58] USDA has not defined *regenerative agriculture*, but it is generally understood to be a holistic suite of practices to restore soil health and ecosystems often based on traditional Indigenous practices, such as Three Sisters (corn, beans, and squash) companion planting.[59]

Farm

The USDA National Agricultural Statistics Service (NASS) defines a farm as any place that produced and sold—or normally would have produced and sold—at least $1,000 of agricultural products during a given year.[60]

Food Security

Food security is the condition "when all people, at all times, have physical and economic access to sufficient, safe and nutritious food that meets their dietary needs and food preferences for an active and healthy life."[61] Community food security extends this to a "condition in which all community residents obtain a safe, culturally acceptable, nutritionally adequate diet through a sustainable food system that maximizes community self-reliance, social justice, and democratic decision-making."[62]

Food Supply Chain

A food supply chain is the network of people, businesses, and processes involved in creating raw food and food products and delivering them to consumers: from farm to fork. It begins at the farm or ranch where the food is sourced and continues through packing, processing, and distribution. It ends with consumption—or eating the food.

Foodshed

A foodshed is the spatial relationships between where food is produced and where it ultimately is consumed. Similar to a watershed, it describes a geographic relationship between an area that supplies a population with food and the flow of food from its origin to destination.

Food Systems

Food systems are multidimensional systems that connect the ecological, economic, and social dimensions of food through feedback mechanisms.[63] Also called food and farming systems, the entire supply chain from production through consumption. *Sustainable food systems* embody the entire life cycle of food—from soil to soil. Building on the principles of sustainable agriculture, they give equal attention to economic, environmental, and social concerns throughout the supply chain and address disposal of food and farm waste.[64]

Land Trust

Land trusts are legal entities—generally nonprofit corporations—that manage land for conservation, affordable housing, and other public purposes. Most are *conservation land trusts* formed to permanently protect environmental resources, like agricultural land, from development. Others are *community land trusts* that hold and steward land for affordable housing and other local assets like community gardens.

Local Food

Local food is an ambiguous term with various definitions. It conjures a sense of place and values, including authenticity, transparency, farm identity, and relationships between producers and consumers. For the purposes of this book, it refers to short, value-based supply chains and direct marketing arrangements, including *direct-to-consumer* (DTC) *sales* through channels like farmers markets and *intermediated sales* to institutions within states or regions. In addition, it includes activities where farmers perform *value-added* functions like packaging, processing, distribution, and promotion.[65]

Planning

Public-sector planning is an international profession and practice. In the U.S., it is a dynamic process to maximize the health, safety, and economic well-being of people in communities. It occurs at many jurisdictional levels, from cities and towns to counties, states, and regions, and encompasses a wide range of specialties from transportation and housing to economic development, hazard mitigation, historic preservation, and sustainability. Planners work in public-sector positions, private and nonprofit sectors, and in volunteer capacities on planning boards and commissions. Food Systems Planning is a set of interconnected, future-oriented activities that support food systems development.

Resilience

Resilience is both the process and the result of successfully adapting to difficult or challenging situations and experiences. It is the ability to prepare for and respond to hazardous events and disturbances and to withstand and bounce back from adversity.

Sustainability

Sustainability is the ability to meet current needs without compromising the needs of future generations. It balances economic growth, environmental care, and social well-being.

How to Use This Book

The book is organized into two sections. Section 1 lays the foundation for planning practices and policy development. Chapter 1 describes the transformation that occurred in U.S. food systems during the second half of the 20th century and explores opportunities for change. Chapter 2 introduces the structures of government that affect planning and how public decisions are made. It covers planning authority and common types of planning organizations and plans that can play a role in food systems planning, pointing to successful examples. Chapter 3 describes principles and practices to apply when planning for farms and food, whether as stand-alone plans or as elements of other types of plans. Building on lessons learned from a variety of planning efforts, it offers basic principles and approaches to guide future planning practices.

Section II focuses on implementation. It is organized around three main elements of food systems planning: land use, sustaining agriculture, and food security. Chapter 4 provides an overview of key federal agencies and policies that influence food systems and, because of their significant roles, goes into greater depth about USDA and the Farm Bill. Chapter 5 gives an overview of state and local land use policies and discusses a variety of zoning considerations that have an impact on food and agriculture. Chapter 6 covers programs and policies to sustain farms and farmland, including those that support agricultural viability, protect farmland from development, and promote environmentally sound farming practices. Chapter 7 focuses on ways to address community food security and health disparities by increasing the availability of fresh fruits, vegetables, and other nutritious foods—from healthy retail policies to nutrition education to emergency food programs. Chapter 8 examines three issues that affect food production but have not received much planning attention: Farmland access, solar energy siting, and commercial cannabis cultivation. While policy responses are evolving, these issues have yet to be included in food systems plans. It also offers some concluding thoughts.

Notes

1. Grant Suneson, "What Are the 25 Lowest Paying Jobs in the US? Women Usually Hold Them," *USA TODAY*, accessed February 2, 2023.
2. Select Subcommittee on the Coronavirus Crisis Majority Staff, "Coronavirus Infections and Deaths Among Meatpacking Workers at Top Five Companies Were Nearly Three Times Higher Than Previous Estimates," *Congress of the United States, House of Representatives*, October 27, 2021.
3. Mike Troy, "Pandemic-Fueled Record Growth in 2020: The PG 100," *Progressive Grocer*, May 17, 2021; USDA, Economic Research Service, "Food Prices and Spending," Updated June 2, 2021; Megan Boyanton, "Pandemic Meat Shortage

Spurs Calls to Shift Slaughterhouse Rules," October 19, 2020; Lillianna Byington, "By the Numbers: Examining the Cost of the Pandemic on the Meat Industry," November 19, 2020; Molly Kinder, Laura Stateler, and Julia Du, "Windfall Profits and Deadly Risks," *Brookings Institution*, November 2020.
4. USDA Economic Research Service, "USDA ERS – Farming and Farm Income."
5. USDA Economic Research Service, "USDA ERS – Food Prices and Spending," January 6, 2023.
6. USDA Economic Research Service, "USDA ERS – Food Security in the U.S.: Key Statistics & Graphics."
7. USDA Food and Nutrition Service, "Barriers That Constrain the Adequacy of Supplemental Nutrition Assistance Program (SNAP) Allotments (Summary)," *USDA Food and Nutrition Service*, June 2021.
8. USDA Economic Research Service, "USDA ERS – Food Security and Nutrition Assistance," October 18, 2022.
9. Sara Usha Maillacheruvu, *The Historical Determinants of Food Insecurity in Native Communities | Center on Budget and Policy Priorities* (Washington, DC: Center on Budget and Policy Priorities, October 4, 2022).
10. Mayuree Rao et al., "Do Healthier Foods and Diet Patterns Cost More than Less Healthy Options? A Systematic Review and Meta-Analysis," *BMJ Open* 3, no. 12 (December 1, 2013): e004277.
11. Fred Magdoff and Harold van Es, *Building Soils for Better Crops*, 4th ed., Sustainable Agriculture Network Handbook Series, Book 10 (College Park, MD: Sustainable Agriculture Research & Education Program, 2021).
12. USDA Economic Research Service, "USDA ERS – Feed Grains Sector at a Glance."
13. D. Pimentel, "Ethanol Fuels: Energy Balance, Economics, and Environmental Impacts Are Negative," *Natural Resources Research* 12 (2003): 127–34.
14. Magdoff and van Es, *Building Soils for Better Crops*.
15. Christine Whitt, Noah Miller, and Ryan Olver, "America's Farms and Ranches at a Glance: 2022 Edition," *USDA Economic Research Service*, December 2022.
16. Whitt et al., "America's Farms and Ranches at a Glance."
17. Thomas A. Lyson, Robert J. Torres, and Rick Welsh, "Scale of Agricultural Production, Civic Engagement, and Community Welfare*," *Social Forces* 80, no. 1 (September 1, 2001): 311–27; Linda M. Lobao, *Locality and Inequality: Farm and Industry Structure and Socioeconomic Condition* (Albany: SUNY Press, 1990); Walter Goldschmidt, *Down on the Farm, New Style* (Yellow Springs, OH: Publisher Not Identified, 1948).
18. Calvin Beale, "Salient Features of the Demography of American Agriculture," in *The Demography of Rural Life*, Publication #64, eds. David Brown et al. (University Park, PA: Northeast Regional Center for Rural Development, 1993); Fred Kirschenmann et al., "Why Worry about the Agriculture of the Middle?," in *Food and the Mid-Level Farm*, eds. Thomas A. Lyson, G. W. Stevenson, and Rick Welsh (Cambridge: The MIT Press, 2008), 3–20.
19. James M. MacDonald, Robert A. Hoppe, and Doris Newton, "Three Decades of Consolidation in U.S. Agriculture," *USDA Economic Research Service*, March 2018.
20. USDA Economic Research Service, "USDA ERS – Farming and Farm Income."
21. 2022 Census of Agriculture Highlights: "Farm Economics." *USDA National Agricultural Statistics Service*, 2024.

22. U.S. Census Bureau, "1920 Census: Volume 5. Agriculture, Reports for States, Chapter 5, Farm Statistics by Race, Nativity, and Sex of Farmer," 1920.
23. 2022 Census of Agriculture Highlights: "Farm Producers." *USDA National Agricultural Statistics Service*, 2024.
24. Eric Jensen et al., "The Chance That Two People Chosen at Random Are of Different Race or Ethnicity Groups Has Increased Since 2010," *Census.gov*, August 12, 2021.
25. Redlining refers to racial discrimination in housing and insurance, originally based on government maps that outlined in red areas deemed risky for investment due to residents' racial characteristics. Heirs property is land jointly owned by family members of a person who died but whose estate did not clear probate. Heirs may use the property but lack clear title, making it hard to obtain loans and federal farm benefits and often forcing partition sales by third parties. Fractionation is similar but refers to land held by Tribes subject to the General Allotment Act of 1887, where reservation land was divided up and allotted to individual Tribal members. After the death of the original allottee owner, title ownership is divided up among heirs. Over generations, the number of owners can grow exponentially, resulting in highly fractionated ownership of much Tribal land.
26. Nina Lakhani, Aliya Uteuova, and Alvin Chang, "The Illusion of Choice: Five Stats That Expose America's Food Monopoly Crisis," *The Guardian*, July 18, 2021, sec. Environment.
27. Food Industry Editorial Team, "Who Are the Top 10 Grocers in the United States?," FoodIndustry.Com, June 7, 2022.
28. Mary Hendrickson et al., "Consolidation in Food Retailing and Dairy," *British Food Journal* 103, no. 10 (January 1, 2001): 715–28.
29. USDA Agricultural Marketing Services, "Packers and Stockyards Division Annual Report 2019" (USDA Agricultural Marketing Service, 2019); Lisa Held, "Just a Few Companies Control the Meat Industry. Can a New Approach to Monopolies Level the Playing Field? | Civil Eats," *Civil Eats*; James M. MacDonald et al., "Concentration and Competition in U.S. Agribusiness," *USDA Economic Research Service*, EIB-256, June 2023.
30. Suneson, "What Are the 25 Lowest Paying Jobs in the US?"
31. Julia Freedgood and Jessica Fydenkevez, *Growing Local: A Community Guide to Planning for Agriculture and Food Systems* (Northampton, MA: American Farmland Trust, April 30, 2017); Cees Leeuwis, Birgit K. Boogaard, and Kwesi Atta-Krah, "How Food Systems Change (or Not): Governance Implications for System Transformation Processes," *Food Security* 13, no. 4 (August 1, 2021): 761–80.
32. Freedgood and Fydenkevez, *Growing Local*; Joachim von Braun et al., "Food System Concepts and Definitions for Science and Political Action," *Nature Food* 2, no. 10 (October 2021): 748–50.
33. Kameshwari Pothukuchi and Jerome L. Kaufman, "The Food System: A Stranger to the Planning Field," *Journal of the American Planning Association* 66, no. 2 (June 30, 2000): 113–24; Kameshwari Pothukuchi and Jerome L. Kaufman, "Placing the Food System on the Urban Agenda: The Role of Municipal Institutions in Food Systems Planning," *Agriculture and Human Values* 16, no. 2 (June 1, 1999): 213–24.
34. Lesli Hoey et al., "Participatory State and Regional Food System Plans and Charters in the U.S.: A Summary of Trends and National Directory," *Michigan State University's Center for Regional Food Systems*, August 2021.

35. World Population Review, "Arable Land by Country 2023."
36. A. Ann Sorenson et al., *Farms Under Threat: The State of America's Farmland* (Washington, DC: American Farmland Trust, 2018).
37. Ann Sorenson et al., *Farms Under Threat*.
38. American Farmland Trust, *Greener Fields: California Communities Combating Climate Change* (Washington, DC: American Farmland Trust, September, 2018).
39. United Nations Department of Economic and Social Affairs, Population Division, "World Population Prospects 2022: Summary of Results," *United Nations Department of Economic and Social Affairs, Population Division*, June 11, 2022.
40. Hans-O. Pörtner et al., eds., "Summary for Policy Makers," in *Climate Change 2022: Impacts, Adaptation and Vulnerability. Contribution of Working Group II to the Sixth Assessment Report of the Intergovernmental Panel on Climate Change* (Cambridge and New York: Cambridge University Press, 2022), 3–33.
41. Jerry M. Melillo, Gary Wynn Yohe, and Terese Richmond, *Climate Change Impacts in the United States: The Third National Climate Assessment* 10.7930/J0z31wj2 OA.Mg (Washington, DC: U.S. Global Change Research Program, 2014); Jeffrey A. Hicke et al., "Chapter 14: North America," in *Climate Change 2022: Impacts, Adaptation and Vulnerability. Contribution of Working Group II to the Sixth Assessment Report of the Intergovernmental Panel on Climate Change* (Cambridge and New York: Cambridge University Press, 2022), 1929–2042.
42. Copernicus Climate Change Service, "Global Climate Highlights 2022 | Copernicus," 2022.
43. Gerald C. Nelson et al., "Climate Change: Impact on Agriculture and Costs of Adaptation."
44. Samuel S. Myers et al., "Increasing CO_2 Threatens Human Nutrition," *Nature* 510, no. 7503 (June 2014): 139–42.
45. Charles L. Walthal et al., *Climate Change and Agriculture in the United States: Effects and Adaptation*, USDA Technical Bulletin 1935 (Washington, DC: USDA ARS, 2012).
46. United States Environmental Protection Agency, "Agriculture," in *Inventory of U.S. Greenhouse Gas Emissions and Sinks: 1990–2020*, 2022, 5–60.
47. Christopher L. Weber and H. Scott Matthews, "Food-Miles and the Relative Climate Impacts of Food Choices in the United States," *Environmental Science & Technology* 42, no. 10 (May 15, 2008): 3508–13.
48. USDA Natural Resources Conservation Service, "Conservation Practices on Cultivated Cropland: A Comparison of CEAP I and CEAP II Survey Data and Modeling – Summary of Findings," *USDA Natural Resources Conservation Service*, January 2022.
49. Keith Paustian et al., "Climate-Smart Soils," *Nature* 532, no. 7597 (April 2016): 49–57.
50. Hans-O. Pörtner et al., "Summary for Policy Makers."
51. USDA Economic Research Service, "USDA ERS – Farm Bill Spending," accessed February 3, 2023.
52. Jill K. Clark, Brian Conley, and Samina Raja, "Essential, Fragile, and Invisible Community Food Infrastructure: The Role of Urban Governments in the United States," *Food Policy, Urban Food Policies for a Sustainable and Just Future* 103 (August 1, 2021): 102014.
53. Monica Hake, Emily Engelhard, and Adam Dewey, "Map the Meal Gap 2022: An Analysis of County and Congressional District Food Insecurity and County Food Cost in the United States in 2020," *Feeding America*, 2022.

54. John Cromartie and Shawn Bucholtz, "USDA ERS – Defining the 'Rural' in Rural America," *Amber Waves*, June 1, 2008.
55. USDA National Agricultural Statistics Service, *2012 Census of Agriculture Highlights: Direct Farm Sales of Food* (Washington, DC: USDA National Agricultural Statistics Service, December 2016).
56. Sections 9–23 of the United States 2022 Census of Agriculture Form Number: 22-A100 provide detail on the types of crops and livestock.
57. USDA National Organic Program Agricultural Marketing Service, "Introduction to Organic Practices," *The NOP Organic Insider*, September 2015.
58. "7 U.S. Code § 3103 – Definitions," *LII/Legal Information Institute*.
59. Emily J. Cole, *Regenerative Agriculture for New England: Sustaining Farmland Productivity in a Changing Climate* (Northampton, MA and Boston, MA: American Farmland Trust and Conservation Law Foundation, August 8, 2022).
60. USDA National Agricultural Statistics Service, "2017 U.S. Census of Agriculture, Introduction," *USDA National Agricultural Statistics Service*, 2017.
61. FAO Agricultural and Development Economics Division, "Food Security (Policy Brief: Issue 2)," *FAO Agricultural and Development Economics Division*, June 2006.
62. Anne C. Bellows and Michael W. Hamm, "U.S.-Based Community Food Security: Influences, Practice, Debate," *Journal for the Study of Food and Society* 6, no. 1 (March 1, 2002): 31–44.
63. Danielle M. Tendall et al., "Food System Resilience: Defining the Concept," *Global Food Security* 6 (October 1, 2015): 17–23.
64. Sonja Brodt et al., "Sustainable Agriculture," *Nature Education Knowledge* 3, no. 10 (2011): 1; C. Clare Hinrichs and Thomas A. Lyson, eds., *Remaking the North American Food System* (Lincoln: University of Nebraska Press, 2008); "Foresight 2.0 – Global Panel," May 31, 2019.
65. Steve Martinez et al., *Local Food Systems: Concepts, Impacts, and Issues* (ERR 97) (Washington, DC: USDA Economic Research Service, May 2010).

1

WHY PLAN FOR FOOD AND AGRICULTURE?

Chapter Summary

This chapter explores the changes to U.S. food systems that occurred during the second half of the 20th century. These changes include:

- A shift to large, specialized, and increasingly consolidated farms and food industries, which led to significant increases in productivity and efficiencies;
- A smaller share of household income spent on food and increased consumption of highly processed foods; and
- Increasing reliance on global trade and supply chains, which both bolstered large-farm profits and increased the variety of foods.

However positive, these changes came at the expense of environmental quality, diet-related disease, and diversified family farms, which—as Covid has shown—leave our current food systems vulnerable to disruptions and shocks. To address these unintended consequences, better food systems planning is needed to increase sustainability, resilience, and fair play throughout the food supply chain.

***

Until Covid-19, most planners and policymakers took for granted that a wide variety of foods from all over the world would be available to most Americans at an affordable price. When the pandemic hit, the U.S. was unprepared for disruptions all along the food supply chain and the sudden surge in home cooking. While we have emergency systems to prepare for and respond to natural disasters and address biohazards and contamination, emergency preparedness plans generally approach these as relatively isolated, short-term events. This was exacerbated by siloed administrative responsibilities and lack of public service coordination.[1] Although this exposed instability in our dominant food systems, it also revealed opportunities for innovation and change.

Places with established local and regional food systems (LRFS) pivoted quickly to fill gaps and leverage community networks to meet production, processing, and marketing needs. LRFS were quick to react, adapt, and innovate, filling gaps in food supply chains at home and abroad.[2]

Food hubs adopted elements of Community Supported Agriculture (CSA) and shifted markets from intermediaries to consumers. Even the federal government adapted aspects of the CSA model and offered funds to distributors to create produce boxes and deliver them to community organizations to distribute to families in need.[3]

Since then the world has faced what the World Economic Forum describes as a "polycrisis": a cluster of weather-related crop failures, war-induced food and energy shortages, and import dilemmas with compounding effects."[4] Predicting what will come if humanity does not take swift and urgent action to mitigate climate change, the Intergovernmental Panel on Climate Change (IPCC) predicts shifts in food production around the world and calls for resilience in food and agricultural systems.[5]

Farmers and ranchers have always been at the mercy of unpredictable weather, but over the past 40 years the extremity of those events has increased in an unprecedented way. Changes in temperature and precipitation have increased the risk of food- and water-borne diseases, respiratory distress from wildfire smoke, and food and nutritional insecurity. They have reduced agricultural productivity, especially in southern and drought-prone regions.[6]

Mega-droughts leave fallowed fields, withered crops, and livestock sold for lack of feed. Scorching heat creates dangerous working conditions. Wildfires destroy crops and livestock, ravage communities, and reduce air quality for thousands of miles. Pressures increase from pests and pathogens and affect when plants and trees bloom and bees and butterflies come out to pollinate them. Heavy rains and floods also harm crops and reduce yields by depleting soil nutrients and increasing erosion. Runoff into waterways diminishes water quality and leads to depleted oxygen levels—or hypoxia—killing fish and shellfish. By mid-century, these impacts are projected to have significant consequences for food security in the U.S. and around the world.[7]

Going forward, we need to support more sustainable, resilient, and evenhanded food systems with sufficient flexibility, diversity, and redundancy to deter, withstand, and recover from individual disruptions as well as a polycrisis. Sustainable because what we do today should meet current needs without compromising future generations. Resilient because we must be prepared to respond to disruptions and change. And evenhanded because communities are healthier when policies are fair and support the needs of all their residents.

How We Got Here

Twentieth-century "cheap food" policies succeeded in holding food prices below the competitive equilibrium price of supply and demand,[8] achieving huge gains in productivity through specialization, consolidation, and use of fossil fuels. Significant investment, both public and private, in intermediate inputs like chemical fertilizers and pesticides, and capital inputs like machinery and farm structures, led to a surge in technological advancement and innovations in animal and crop genetics. Total domestic farm output nearly tripled between 1948 and 2019[9] and, combined with liberalized global trade policies, increased food availability at lower costs (see Figure 1.1). However, agricultural growth has slowed in recent years, with wide variations in annual output largely attributed to adverse weather.[10]

Over this period, the use of farm inputs was fairly steady but its composition changed. Inputs like labor and land declined significantly—by 74 percent and 28 percent, respectively—while intermediate inputs grew by 126 percent and capital inputs (excluding land) increased by 79 percent.[11] Agriculture became increasingly specialized, consolidated, and focused on

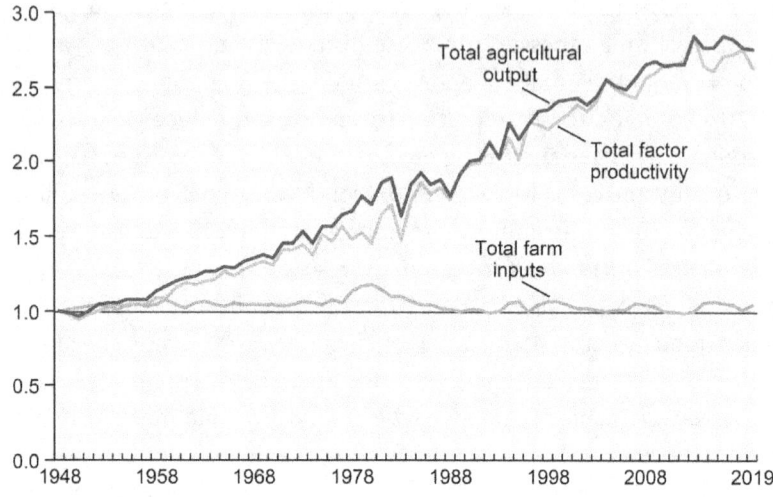

Figure 1.1 Total domestic farm output nearly tripled between 1948 and 2019.

a few highly productive commodity crops well suited to industrial farming systems—like corn. Grown primarily for livestock feed and ethanol, corn is the largest crop in the U.S., accounting for over 95 percent of total feed grain production and use, and our largest export crop.[12] But this dramatic shift came with significant environmental costs from degraded soils to hypoxia in the Gulf of Mexico.[13]

Consolidation—and increasingly concentration—also occurred in agricultural input industries and markets. Sometimes resulting from a desire to achieve economies of scale, consolidation can increase productivity. But when it becomes highly concentrated into only a few firms, it can slow productivity growth, reducing competition, efficiency, and innovation.

Today just two seed companies account for 72 percent of planted corn acres. Over 30 years, as the number of companies shrank, the prices farmers paid for seed increased by an average of 270 percent—far outstripping commodity output prices. Meanwhile, meatpacking shifted to larger plants and tighter vertical coordination among production and processing. By 2019, four meatpackers accounted for 85 percent of beef and 67 percent of hog slaughter.[14] During the pandemic, it became clear how vulnerable this kind of concentration is to disruptions and sudden closures. Early on, Covid outbreaks shuttered 25 percent of pork processing.[15]

Large packing plants achieve cost savings through economies of scale. But these are modest—as little as a one to three percent cost advantage over smaller plants.[16] And without competition, they can collude to keep wages low and food prices high. Over the past few years, a series of lawsuits have been settled against meat processing companies for fixing prices and conspiring to depress compensation.[17] Systems with greater diversity of scales disperse the risk of any one plant closing and improve market access for sustainable livestock production.

Finally, since they only achieve economies of scale when operating at or near full capacity, consolidated meat packing has contributed to the shift toward large, concentrated animal feeding operations (CAFOs). These create severe environmental problems from air and water pollution to GHG emissions, not to mention animal welfare concerns. While regulations are in place, they are riddled with exemptions and loopholes, so some of their true costs are passed on to society.[18]

On the retail end, four transnational firms control at least half the market for most common groceries.[19] In most metropolitan areas, five to six chain stores account for most supermarket sales.[20] Walmart alone sells at least half of all groceries in one in every ten metropolitan areas and nearly one in three "micropolitan" areas.[21] These trends reduce marketing outlets for small and midsize producers and food processors, and give big processors and retailers more leverage and power to dictate prices to suppliers and consumers.

Twentieth-century policies also led to a loss of demographic diversity in agriculture. In the 1920 Census, nearly 15 percent of U.S. farmers were reported as "Negroes," "Indians," "Japanese," or "Chinese."[22] Today, "White" farmers represent over 95 percent of all farmers—in striking contrast to the overall U.S. population, which is increasingly multiracial and diverse.[23]

As these changes occurred in agriculture, Americans' diets increasingly shifted from home-cooked meals created from raw ingredients to reliance on processed foods and take-out. With this, the share of the food dollar shifted from farmers to supply chain intermediaries. In 2021, U.S. farms received only 14.5 cents of every dollar spent on domestically produced food; the rest went to the supply chain. (Note, the farm share jumped slightly in 2020 due to a sharp decrease in spending on food-away-from-home early in the Covid-19 pandemic.)[24] (See Figure 1.2.)

By 2021, disposable income spent on food comprised only about 11 percent of household budgets,[25] even though Americans have significantly increased food consumption. A Pew Research Center study found that an

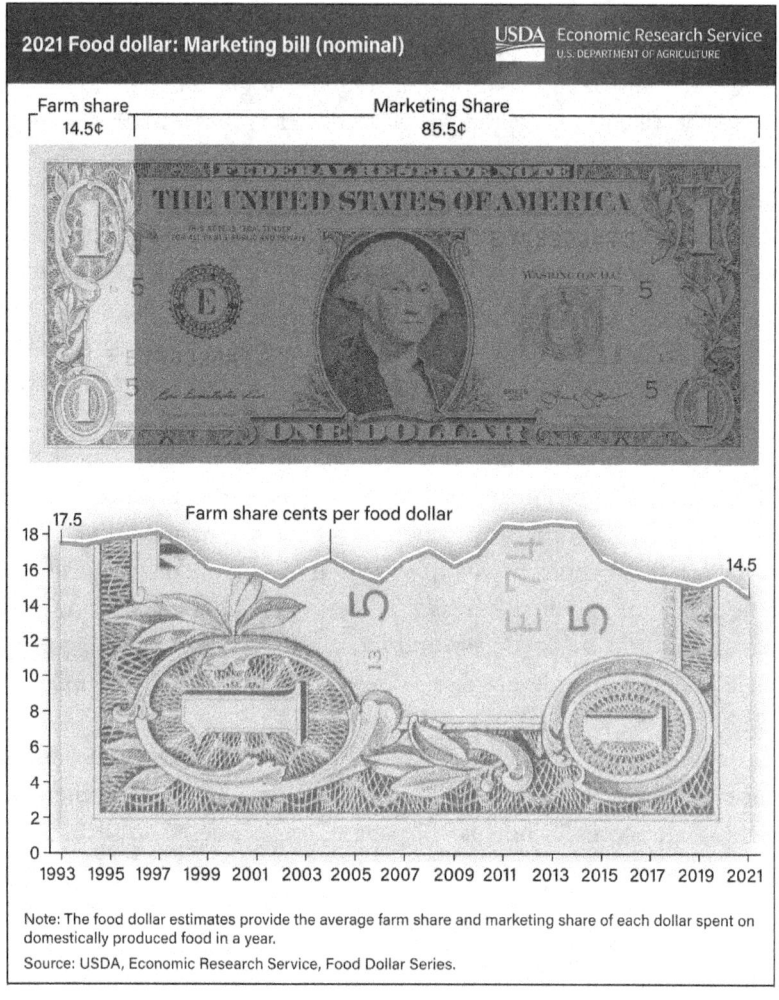

Figure 1.2 ERS uses input-output analysis to calculate the farm and marketing shares from a typical food dollar. In 2021, U.S. farms received only 14.5 cents of every dollar spent on domestically produced food.

average American consumed about 23 percent more calories a day in 2010 than in 1970.[26]

Considering widespread availability and consumer price, cheap food policies achieved their goals. But they came with unintended consequences: the decline of diverse family farms and rural economies, food insecurity and diet-related disease, and natural resource degradation.[27]

Decline of Diverse Family Farms and Rural Economies

As farm output rose, farms got larger and farm numbers fell. According to the Census of Agriculture, in 1950, 5.4 million farms averaged 215 acres. By 1997, average farm size had doubled and farm numbers fallen to 2.2 million. The decline was so dramatic that USDA described it as a free-fall situation leading toward trauma.[28] The drive for large, technology-dependent, specialized farms squeezed small commercial and midsized producers and led to substantial consolidation in farm sales.

The free fall disproportionately affected farmers of color. This has been especially well documented for Black/African American (BAA) farmers. Between 1920 and 1997, they lost 90 percent of their land—a resource valued at roughly $326 billion.[29] Several audits, investigations, and lawsuits found that USDA county committees denied loans, held up loan approvals, and offered worse terms to farmers of color than to White farmers, forcing many into high debt and foreclosure—even after the 1964 Civil Rights Act was supposed to end racial discrimination in federal programs. Forced partition sales from heirs property exacerbated the problem, contributing to a massive transfer of land and wealth to White farmers. Since 1910, White farmers increased farmland ownership by 89 percent, while BAA farmers decreased ownership by 80 percent.[30]

Two class-action law suits: *Pigford vs. Glickman* and *Brewington vs. Glickman* (which were consolidated), established that USDA had systematically discriminated on the basis of race. Settled in 1999 for nearly $1 billion, *Pigford I* was the largest civil rights settlement in history and set a precedent for future class-action law against USDA. In 2008, Congress directed that all pending claims and class actions brought against USDA by socially disadvantaged farmers or ranchers, including Hispanics and women, be resolved in an expeditious and just manner and established a $1.33 billion settlement fund to resolve pending claims. In 2010, USDA settled *Pigford II* for a record

$1.25 billion because so many eligible BAA farmers had been left out of the first settlement. It also settled *Keepseagle v. Vilsack*, a long-running class-action suit that awarded a $680 million to eligible Native Americans, plus loan forgiveness and tax relief. *Keepseagle* also established a Native American Agriculture Fund to provide business assistance, agricultural education, technical support, and advocacy services.

Today, consolidation persists with a continued shift of production to the largest farms. Farm size keeps creeping up while farm numbers decline, just not so dramatically. In 2021, 5 percent of large-scale family and nonfamily farms supplied 63 percent of production—up from 31 percent in 1991 (adjusted for inflation).[31] Cropland also is increasingly concentrated on fewer, larger farms. In 1987, farms operating at least 2,000 acres accounted for 15 percent of all cropland. Their share more than doubled to 37 percent in 2017, and the share of crops from farms with at least 10,000 cropland acres quadrupled.[32]

Still, small and midsized family farms—where most of the business is owned by operators and individuals related to them—comprise 95 percent of all operations and manage nearly two-thirds of America's agricultural land. They include many kinds of operations, from retirement farms to small farms whose principal operators report farming as their primary occupation. They also include most local-food farms and farms operated by young, beginning, and historically disadvantaged producers. Not counting retirement farms, which represent 11 percent of the total but only 1 percent of production, small and midsized farms contribute 35 percent of the value of production.[33] Squeezed between large, consolidated farms and concentrated input industries, processors, and markets, they struggle economically and need more and better support from planners and policymakers (see Figure 1.3).

Often the backbone of rural communities, these farms have long been associated with a civically engaged middle class and more equitable distribution of income, supporting cohesive communities.[34] They tend to be innovative, flexible, and produce differentiated products. They buy livestock, equipment, and supplies locally, and employ local people. The decline of these farms has hollowed out rural communities and reduced the diversity of supply chains.[35] Declining employment and economic opportunity has led to rural depopulation and the loss of essential social infrastructure such as hospitals, schools, and banks.[36]

WHY PLAN FOR FOOD AND AGRICULTURE?

Distribution of farms, land operated, and value of production by farm type, 2021

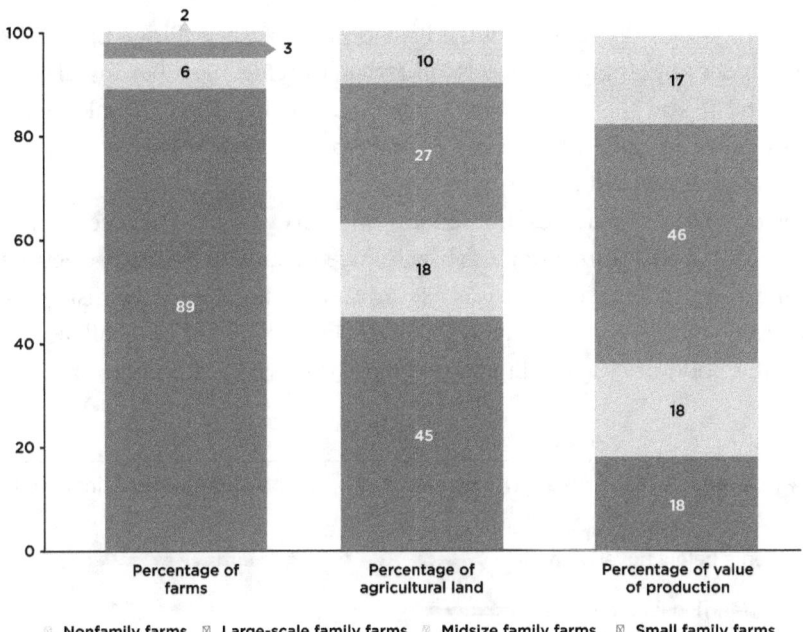

Figure 1.3 Small and midsized family farms are squeezed economically between large, consolidated farms and an increasingly concentrated supply chain.

Food Insecurity and Diet-Related Disease

The rapid suburbanization that occurred after World War II moved significant wealth to the suburbs. Combined with agriculture's dramatic restructure, urban disinvestment, and discriminatory redlining polices, it led to concentrated poverty in both rural communities and the inner city, where today people experience the highest degrees of food insecurity.

In 2021, 32.1 percent of households with incomes below the federal poverty line were food insecure—meaning they lacked access, at all times, to enough food for an active, healthy life for all household members. Most prevalent in large cities and rural areas, it affected 12.5 percent of households with children, with substantially higher rates for single-mother, Black, and Hispanic households[37] (see Figure 1.4).

Even with the Supplemental Nutrition Assistance Program (SNAP), 88 percent of participants reported barriers to achieving a healthy diet, and 61 percent said they could not afford to buy healthy food. Coupled with a lack of cooking equipment and/or storage, affordability was most associated with household food insecurity.[38]

Food insecurity forces households to make choices between buying nutritious food and meeting other basic needs, such as housing, electricity, health care, and childcare. It was exacerbated by the economic disruptions that came with Covid-19. Supply chain disruptions and other challenges led to rising food prices and illuminated racial disparities and policy failures.[39]

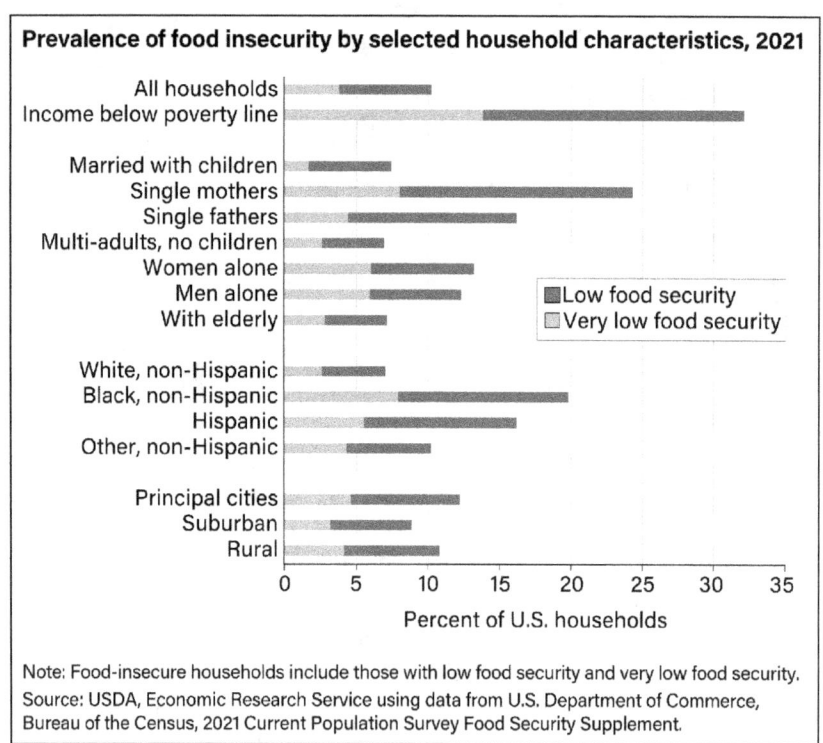

Figure 1.4 Food insecurity rates are highest for single-mother households and households with incomes below the poverty line.

The federal response helped keep hunger at bay but was temporary. Since then, global unrest and extreme weather events have disrupted supply chains and, combined with market concentration, likely will maintain inflationary food prices.

While food security is associated with access to affordable food, access alone cannot ensure the food is nutritional or will lead to an active, healthy life for all household members. Nutritional security prioritizes wholesome food that leads to good health. It is important for all members of society. Yet nearly half of Americans' calories now come from processed foods like flours, fats, and oils, with rising household expenditures on these and foods like frozen and packaged meals. Dairy, meat, fruits, and vegetables comprise a smaller share today than 40 years ago.[40] Using a machine-learning algorithm to predict the degree of processing for any food, a 2023 study found more than 73 percent of the U.S. food supply is ultra-processed and correlated this with higher risk of chronic diseases, including diabetes, angina, and elevated blood pressure.[41]

Mounting evidence links ultra-processed foods to poor nutritional quality, rising rates of obesity, and chronic diseases.[42] Indeed, the prevalence of obesity in the U.S. more than tripled from 1960 to 2020.[43] Ultra-processed foods are cheaper, faster, and more convenient than cooking meals with whole foods and ingredients. A systematic review and meta-analysis of the relationships between the healthfulness of foods/diet patterns and their price, found healthier patterns were more expensive, creating difficulties for many low-income families. It also found the price difference trivial in comparison to the costs of diet-related chronic diseases.[44]

Natural Resource Degradation

Since Rachel Carson exposed the hazards of the pesticide DDT in *Silent Spring*, evidence has mounted about the ecological problems of the "conventional" agriculture common after World War II. The focus on productivity and specialization significantly increased yields, but did so with increasing use of fossil fuels and petrochemical inputs. Combined with geographic separation between crop and livestock production, it resulted in a host of negative environmental impacts: degraded soil and water quality, elevated GHG emissions, and loss of biodiversity, which have been widely reported.[45]

Figure 1.5 The Red River Basin is located in parts of Minnesota, North Dakota, and South Dakota. Erosion has caused severe flooding, which has resulted in loss of wildlife habitat and increased water quality concerns.
Source: USDA. Photo by Keith Weston.

On the crop side, specialization led to dependence on synthetic fertilizers and pesticides, which run off into waterways. On the livestock side, it led to concentrated animal feeding operations (CAFOs). These require massive manure lagoons, which leak contaminants during extreme rain events and/or due to poor construction. Water pollution from nitrogen and phosphorus runoff have had significant regional effects, as in the case of the Chesapeake Bay, and continental impacts, as evidenced by the hypoxia in the Gulf of Mexico. Antibiotics, commonly used to promote rapid growth in poultry and livestock, also leak into waterways,[46] and their overuse in animal feed ends up in meat, which has led to bacteria resistance to antibiotics used for human health.[47]

Conventional agriculture uses soil as a medium to be tilled and amended rather than a dynamic living resource. Coupled with the use of large equipment, this has resulted in erosion, topsoil loss, and dry, compacted soils with low organic matter, unable to hold water. Further, tilling kills beneficial microorganisms in the soil and contributes to GHG emissions.

Agriculture also contributes to the loss of biodiversity—the variety and variability of life on earth. Biodiversity is vital for growing food and delivering fresh water, thus to human health. Losses can impact nutrition, the availability of traditional medicines, and patterns of infectious diseases. Worldwide, about one million species of animals and plants are threatened with extinction from agricultural land use changes and activities like logging, as well as climate change, pollution, and invasive species.[48]

Finally, according to the U.S. Environmental Protection Agency (EPA), domestic agriculture contributes about 10 percent of total U.S. GHG emissions. Much of this comes from liquid manure from CAFO lagoons, but cows and other ruminants also produce methane as part of their normal digestive processes. In addition, nitrogen fertilizers and nitrogen-fixing crops emit nitrous oxide—which also is released by draining organic soils and certain irrigation practices. Yet at the same time, EPA reports that agricultural land use and forestry have removed more CO_2 from the atmosphere than they have created in emissions—and thus are considered a net sink.[49]

The more farmers and ranchers adopt soil health and other conservation practices, the more they will contribute to a healthy environment and help to combat climate change.

Where We Go From Here

Since 2001, the Environmental Working Group has tracked federal support to agriculture. It found that between 1995 and 2021, USDA spent $478 billion on commodity payments, crop insurance, conservation, and disaster relief. Based largely on acreage or production, most of its payments went to the largest and wealthiest farms. The top 10 percent of largest recipients received nearly 80 percent of federal payments, while 80 percent of the smallest recipients only received about 9 percent.[50]

Large farms achieve economies of scale through specialization and expansion, but small and midsized farms are more likely to succeed by diversifying and expanding their scope. Given that diversified systems have ecological and community benefits, more research, better planning, and a new policy paradigm are needed to support diversification strategies both on the farm and through local and regional supply chains.

Planners are uniquely poised to advance sustainable, resilient, and evenhanded food systems. Increasing productivity and maintaining food affordability remain important, but we must achieve these without compromising the health of our land, our communities, and our people. Planners generally have a good understanding of public programs and processes, and their experience and skills can illuminate the social, economic, and environmental impacts of public decisions. Yet modern conditions call for a fresh look at how to approach planning, including food systems planning.

The impacts of increasingly severe weather events challenge the firm jurisdictional boundaries upon which land use planning and policy rely. As a changing climate further disrupts food production and distribution, we increasingly will need to plan across—rather than within—jurisdictions. Water moves, smoke moves, even soil moves under certain conditions. And of course, food moves along the supply chain, affecting many sectors of the economy and society. Further, while planning as a profession was developed largely to guide land use and development in cities, food systems span a rural–urban continuum. While most specialty crops like fruits, nuts, and vegetables come from metropolitan counties, rural counties supply most of our meat and staple foods like oats, rice, and wheat. Long and short, the complex forces that affect food systems transcend traditional boundaries. They are structural not individual sector issues, and must be addressed together, interwoven with other planning and preparedness activities.

It Starts With Land Use and Land Management

Including public lands, farmers and ranchers manage over a billion acres—more than half of the land base of the contiguous U.S. This abundant and varied landscape includes crop, pasture, range, woodland, and federal grazing lands that supply food, feed, fiber, fuel, and flowers, sustaining millions of jobs and the nation's balance of trade (see Figure 1.6).

Well-managed agricultural lands support ecological services from carbon sequestration to groundwater recharge, erosion and flood control, fire suppression, wildlife and pollinator habitat, scenic views, and recreational opportunities.[51] They also supplement the commercial food supply with resources for hunting, fishing, foraging, and home and community gardens.

Despite their importance to food security, economic prosperity, and environmental quality, since 1982, the U.S. has lost about 26 million acres of farmland to urbanization and poorly planned development—an area larger than the state of Kentucky,[52] with millions more acres fragmented by low-density residential development outside of urban limits, which creates conflicts with neighbors and threatens the viability of agricultural economies.[53]

Agriculture is a Dominant Feature of the American Landscape

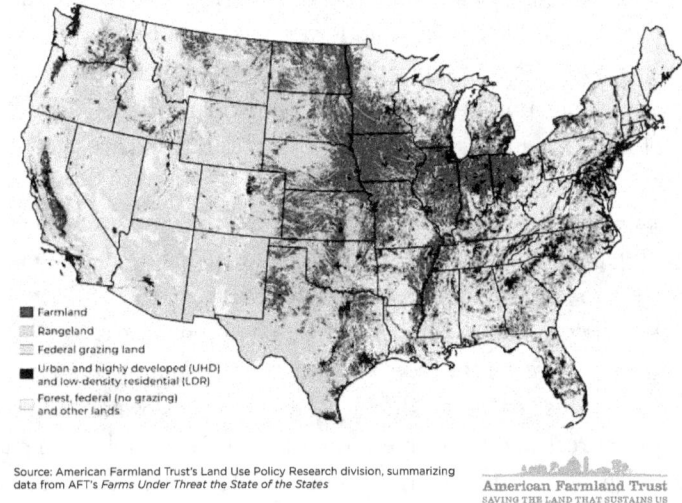

Figure 1.6 Agricultural lands create a mosaic of land covers and uses, including farmland, rangeland, and federal lands used for grazing, juxtaposed against forests, other federal lands, urbanized areas, and low-density residential development.
Source: American Farmland Trust, "Farms Under Threat: The State of the States."

Beyond scattered housing development, high-quality farmland is threatened by forces including warehouse development, utility scale solar energy installations, even environmental preservation.

Based on 20th-century development patterns, planners developed a smart growth toolbox to address urban sprawl. Smart growth is an established approach to planning and development that encourages a mix of building types, uses, housing, and transportation options. It directs development toward existing communities and infrastructure, and balances new construction with the protection of farmland, open space, and other natural areas. Even without coordinated policy effort, urban sprawl has slowed.[54] But large-lot, low-density residential land use still drives development in rural regions, fragmenting the agricultural land base, inflating land values, and paving the way for urbanization.[55] With improved rural internet service and more remote work, without more and better rural planning, this likely will increase. But so far, scattered rural development has not evoked much planning or policy response.

Rural communities also face the impacts of farm consolidation. Once diversified agricultural landscapes become dominated by monoculture, wealth and power shift to fewer and fewer people, populations shrink, and communities lose shopping, health care, and other vital services. As farms get larger, they bid up land prices and remove land from rental markets, making it harder for small and midsized operations to expand and for young and beginning farmers to enter agriculture.

The challenge for smaller operations is exacerbated because the income from farming generally does not support farm real estate values—even when interest rates are low.[56] With cropland values rising sharply and interest rates at a 30-year high, without better land use planning, access to land will be increasingly out of reach for most producers (see Figure 1.7).

States' ambitious goals for reducing GHG emissions by expanding renewable energy sources add to competition for land. Renewables generate electricity through natural and recurring processes. Ideally, they provide financial returns to complement, rather than replace, agricultural production. These come through cost savings from generating energy to support the farm or ranch operation, or from leasing land to a renewable energy developer. However, transitioning to renewable energy requires dramatic increases in the number and scale of solar and wind installations. This largely will occur

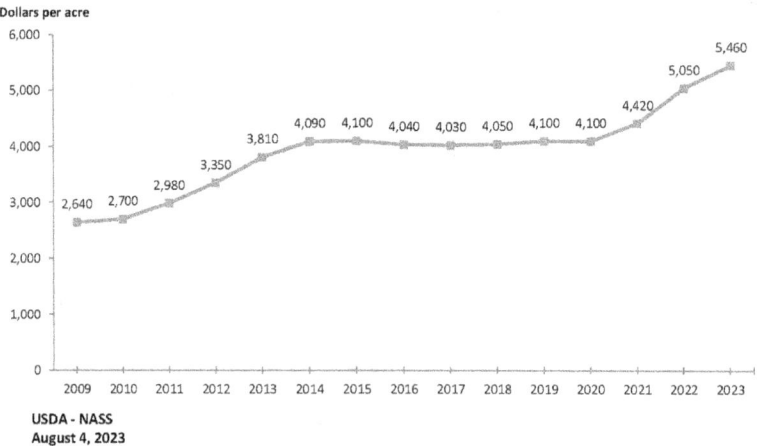

Figure 1.7 Cropland values have risen sharply since 2020.

on rural lands. Without comprehensive planning to balance competing land uses, this threatens to take more land out of agriculture—reducing not only food production but the many environmental benefits that well-managed farmland can provide. This includes carbon sequestration that itself offsets GHG emissions. In our rush to advance renewable energy to meet climate goals, it's essential to remember the value of farmland to meet similar goals.

The U.S. Department of Energy (DOE) predicts that by 2030, 20 percent of U.S. energy needs will be met by wind, increasing to 35 percent in 2050.[57] Most land-based wind generation takes place in nine traditional farm states: Illinois, Iowa, Kansas, Minnesota, Nebraska, North Dakota, Oklahoma, South Dakota, and Texas. Turbines only use a small amount of land so farmers can farm around them. So long as soils are protected during construction, turbines, relay boxes, and other infrastructure should not significantly hamper agricultural production and may even improve row crop yields.[58] However, these critical steps are not always taken because wind energy developers seldom understand agriculture or consider strategies that would support it.

Solar energy also can be compatible with farming. Some farmers and ranchers install solar panels on barns and other structures, in ditches or

Figure 1.8 Windmills on a farm in Audubon County, Iowa.
Source: USDA. Photo by Cecilia Lynch.

marginal fields, to generate energy for their own use, or to sell excess power. This can support their bottom line when the primary use remains agriculture. Further, some farmers are exploring the fast-evolving field of agrivoltaics (or "dual-use"), where solar panels are spaced strategically to allow crops or livestock underneath and between them. However, at present, utility-scale solar installations blanket large swarths of flat, open, and well-drained farmland without regard to agriculture or food production, taking valuable acreage out of production.

Solar developers buy and lease land to install large-scale arrays, which require substantial acreage and often lead to land use conflicts. Extensive steel infrastructure and a sizeable footprint both use and shadow land, causing significant land conversion and limiting farm production. (See Chapter 8 for a more in-depth discussion of solar siting on farmland.) Communities—especially rural communities—need planning support to weigh competing priorities and direct development to disturbed and marginal lands while protecting arable soils. Otherwise, food production and environmental quality will be compromised as cultivation is pushed to poor-quality soils that achieve lower yields, require more petrochemical inputs, and lead to higher GHG emissions, erosion, and pollution.

Healthy soils produce healthy food and support a healthy environment. Farming, when done right, can improve soil health. Yet farming also can harm the environment. Growing global and domestic demands to significantly increase food production could lead more farms to impair soil health. Beyond needing farmers and ranchers to grow food and provide for other basic needs, we need to promote environmentally sound farming practices. This requires attention to land management strategies as well as to land use.

"The nation that destroys its soil destroys itself."
Franklin Delano Roosevelt[59]

Science is going through a paradigm shift when it comes to understanding soil. It once was thought that plants consumed soil through their roots the way that animals eat with their mouths. Plowing—or tilling—helped farmers break soil into tiny bits to "chew" their crops' food. During the "Green Revolution" of the mid-20th century, it was thought that soil absorbs nutrients from a soil solution, which could be boosted with chemical fertilizers. Today, scientists recognize that carbon in the soil is filled with living

Figure 1.9 Healthy soil looks dark, crumbly, and porous, and is home to worms and other organisms.
Source: USDA.

things: organisms that live on carbohydrates from plant roots and other organisms that decompose things. Plants can uptake some nutrients dissolved in a soil solution, but they get most of what they need from the sun and the soil microbial community.

Microorganisms that live in soil work together to support crops as well as animal and human health, water quality, and climate adaptation and mitigation. Plants capture CO_2 through photosynthesis and convert it to organic carbon compounds in plant tissues. Some of this carbon is stored—or sequestered—in the soil.[60] Rebuilding carbon stocks in soil is vital both for farm productivity and to mitigate climate change.

Some farmers already are adopting regenerative—or climate smart—conservation practices. These include no-till, strip-till, and reduced-tillage; cover crops; diversified rotations; intercropping or double cropping; rotational livestock grazing; and incorporating livestock or agroforestry systems.

Especially when adopted as whole-farm systems instead of individual practices, they can stabilize yields while increasing soil health and improving water quality and holding capacity to withstand droughts and flooding.

Different types of production require different types of soils, provide varying degrees of habitat, and rely on other natural assets. Most development—whether renewable energy facilities, affordable housing, or supermarkets—can occur on lesser-quality soils. USDA's Natural Resources Conservation Service (NRCS) has data and maps to guide planning natural resource management. Available for 95 percent of U.S. counties, they can be used in conjunction with its Soil Survey Geographic Database (SSURGO), which contains a century of data on yields for different agricultural uses, water capacity and frequency of flooding, and engineering uses such as limitations on building-site development.

ALL FARMLAND IS NOT CREATED EQUAL

The 1981 Farmland Protection Policy Act defined important farmland to minimize federal contributions to farmland conversion and ensure federal programs are compatible with state, local, and private farmland protection programs and policies. These definitions are often used for other statutory purposes.

- *Prime farmland* has the best combination of physical and chemical characteristics for producing food, feed, forage, fiber, and oilseed crops. It has the best soil quality, growing season, and moisture supply to economically produce sustained high yields when managed according to acceptable farming practices.
- *Unique farmland* is well suited to produce specific high-value crops, such as citrus, tree nuts, olives, fruits, and vegetables. It has a special combination of soil quality, location, growing season, and moisture supply to economically produce sustained high and/or high-quality yields of specific crops when managed with acceptable farming practices.
- *Farmland of statewide importance* is in addition to prime and unique farmlands. Generally, it includes lands that are nearly prime and may produce equivalent yields in favorable conditions. State agencies determine the criteria for defining and delineating these lands, which may include tracts designated for agriculture by state law.

Building on these categories, American Farmland Trust (AFT) created a new framework to assess productivity as well as the versatility and resilience (PVR) of agricultural lands for long-term food production. AFT mapped PVR and identified lands of national significance, as well as each state's best land for edible food crops like fruits, nuts, vegetables, and staple grains. AFT also created web mapping tools to visualize past and projected threats and inform land use decisions.

Factors Used to Determine PVR:

- ***Soil suitability*** uses important farmland categories to assess soil capacity to support agricultural production (productivity) and provides clues to the land's versatility and resiliency to withstand weather extremes.
- ***Land cover/use*** shows where different types of agriculture are practiced. Land cover includes vegetation or other material that covers the surface of the land. Land use is the purpose of human activity on the land.
- ***Food production*** is an important proxy for characteristics that support specialty crop production, which may require unique soils and microclimates. We included 132 individual cropland types divided into five main groups: 1. fruit and nut trees; 2. fruits and vegetables grown as row crops; 3. staple food crops (e.g., wheat, rice, potatoes); 4. feed grains, forages, and crops grown for livestock feed and processed foods; and 5. non-food crops.[61]

It Continues Through the Supply Chain

Disruptions in food supplies threaten food safety, security, and nutrition, especially for vulnerable populations. Beyond land use and land management, we need to build greater sustainability, resilience, and parity into food processing, storing, and distribution systems. Locating more decentralized, lower-scale facilities near consumers would reduce storage and transportation costs and potentially environmental impacts. More medium-sized processors would have access to open markets, which would improve traceability and price transparency but still provide efficiencies and economies of scale. Expanding regional meat processing would encourage more sustainable livestock production, reduce agriculture's GHG emissions, protect small and midsized farmers, and protect consumers from price gouging. Paying food system workers a living wage would relieve food insecurity.

Change is starting to happen. In 2023, USDA announced a dozen Regional Food Business Centers to bolster LRFS and help small and midsized farms and food businesses reach new markets and resources. The centers cover either a 400-mile local food radius or parts of at least three states or territories. USDA has committed to supporting high-need areas with persistent poverty and limited resources to target historically underinvested communities.

Planning at multiple levels and scales can support keeping farmland in farming. It can stimulate the middle infrastructure needed to shorten supply chains and connect producers with processors, distributors, and the growing number of consumers who desire nutritious, traceable food. It can ensure alternatives if transportation routes are blocked, conceive ways to consolidate deliveries to reduce emissions, encourage storage centers with access to markets, and ensure food access for low-income consumers. It can address food waste throughout the supply chain, reducing municipal solid waste and redirecting edible food to food pantries and kitchens. And it can support coordinated emergency response.

Leading the way, in 2021 the Maryland state legislature passed a Maryland Food System Resiliency Council bill to address food insecurity stemming from the pandemic and subsequent economic crisis. The Council is tasked with developing policy recommendations to advance food system equity and sustainability and drafting a plan to increase the production and procurement of Maryland certified food.

Building on Maryland's example, strengthening diversified local and regional food systems is a good place to begin. LRFS supplement rather than supplant dominant food systems with shorter supply chains and a commitment to local communities. Generally associated with sustainable agriculture, they include direct-to-consumer arrangements through farmers' markets, farm stands, and CSAs, as well as intermediated markets like farm to school or restaurants and supermarkets, and hybrid arrangements. These relationships are more customer-focused and responsive than wholesale operations, and as we learned during Covid, producers can pivot quickly to adapt to changing market conditions in the event of disruptions or disasters. They also stimulate local economic activity, generating new jobs and businesses directly and indirectly through food processing, distribution, and related sectors[62] and promote rural economic revitalization.[63] Ideally, this will occur within a larger universe of land use planning, smart growth, life-cycle thinking, emergency preparedness, and so on.

As the adage goes, "A failure to plan is a plan to fail." The policies that led to profound changes in the 20th century must be retooled for the 21st—especially to prepare for the impacts of climate change. We need more and better food systems planning to usher in a 21st-century transformation before it is too late.

Discussion Questions

1. What are the characteristics of agricultural land and land use in your community? (Consider development pressure, land values, types of agricultural production, soil types, etc.)
2. Describe the food system or systems that support your community. Draw a picture or use a map to identify where key parts of your food system take place, from farms to food processors and distributors, to food retail and restaurants, to disposal of food and agricultural waste.
3. Who controls the food supply chains in your community, and what might happen if one or more steps in the supply chain are disrupted?

Notes

1. Shoshanah Inwood et al., *Preparing for Food System Resiliency in Ohio: Policy and Planning Lessons from COVID-19* (Columbus, OH: College of Food, Agricultural and Environmental Sciences & John Glenn College of Public Affairs, January 2022).
2. James Worstell, "Ecological Resilience of Food Systems in Response to the COVID-19 Crisis," *Journal of Agriculture, Food Systems, and Community Development* 9, no. 3 (2020): 23–30; Angelina Sanderson Bellamy et al., "Shaping More Resilient and Just Food Systems: Lessons from the COVID-19 Pandemic," *Ambio* 50 (2021): 782–93; Dawn Thilmany et al., "Local Food Supply Chain Dynamics and Resilience during COVID-19," *Applied Economic Perspectives and Policy* 43, no. 1 (2021): 86–104.
3. Thilmany et al., "Local Food Supply Chain Dynamics and Resilience during COVID-19."
4. World Economic Forum, "We're on the Brink of a 'Polycrisis' – How Worried Should We Be?," January 13, 2023.
5. Pörtner et al., eds., "Summary for Policy Makers."

6. Rachel Bezner Kerr et al., "Food, Fibre, and Other Ecosystem Products," in *Climate Change 2022: Impacts, Adaptation and Vulnerability. Contribution of Working Group II to the Sixth Assessment Report of the Intergovernmental Panel on Climate Change* (Cambridge and New York: Cambridge University Press, 2022), 713–906. doi:10.1017/9781009325844.007; U.S. Global Change Research Program, *Climate Change Impacts in the United States: The Third National Climate Assessment* (Washington, DC: USGCRP, 2014). doi:10.7930/J0Z31WJ2.
7. U.S. Global Change Research Program, *Climate Change Impacts in the United States: The Fourth National Climate Assessment* (Washington, DC: USGCRP, 2018). doi:10.7930/NCA4.2018; M. Rohde Melissa, "Floods and Droughts Are Intensifying Globally," *NatureWater* 1 (2023): 226–27; Bezner Kerr et al., "Food, Fibre, and Other Ecosystem Products."
8. Ronald Knutson et al., *Agricultural and Food Policy*, 4th ed. (Prentice-Hall, 1998).
9. USDA Economic Research Service, "USDA ERS – Farming and Farm Income."
10. Sun Ling Wang et al., "U.S. Agricultural Output Has Grown Slower in Response to Stagnant Productivity Growth," USDA, Economic Research Service, May 2023.
11. Wang et al., "U.S. Agricultural Output Has Grown Slower in Response to Stagnant Productivity Growth."
12. USDA Economic Research Service, "USDA ERS – Feed Grains Sector at a Glance."
13. Pimentel, "Ethanol Fuels," 127–34.
14. MacDonald et al., "Concentration and Competition in U.S. Agribusiness."
15. Amelia Lucas, "Meatpacking Union Says 25% of US Pork Production Hit by Coronavirus Closures," CNBC, April 23, 2020.
16. James M. MacDonald et al., "Consolidation in U.S. Meatpacking," *USDA Economic Research Service*, AER-785, 2000.
17. Josh Funk, "Are You Paying Too Much for Bacon? One of the Big Meat Producers Just Settled a Price-Fixing Lawsuit for $20 Million," *Fortune*, September 20, 2022; Todd Neeley, "$10M Settlement Reached in Wages Case Settlement Struck in Antitrust Case Alleging Food Companies Suppressed Wages," *Progressive Farmer*, August 8, 2023; Mike Scarcella, "Pork Consumers' $75 Million Price-Fixing Accord With Smithfield Approved," April 12, 2023.
18. Lindsay Walton and Kristen King Jaiven, "Regulating CAFOS for the Well-being of Farm Animals, Consumers and the Environment," *Environmental Law Reporter*, 50 ELR 10485, June 2020.
19. Lakhani et al., "The Illusion of Choice."
20. MacDonald et al., "Concentration and Competition in U.S. Agribusiness."
21. Stacy Mitchell, "Rep. Walmart's Monopolization of Local Grocery Markets," *Institute for Local Self-Reliance*, 2019.
22. 1920 Census: Volume 5. Agriculture, Reports for States; 1920 Census. Chapter 5: Farm Statistics by Race, Nativity, and Sex of Farmer.
23. N. Jones, R. Marks, R. Ramirez, and M. Rios-Vargas, "2020 Census Illuminates Racial and Ethnic Composition of the Country," *U.S. Census Bureau*, 2021.
24. USDA Economic Research Service, "Farm Share of U.S. Food Dollar Reached Historic Low in 2021," 2022.
25. USDA Economic Research Service, "USDA ERS – Food Prices and Spending," January 6, 2023.

26. Drew Desilver, "What's on Your Table? How America's Diet Has Changed over the Decades," *Pew Research Center*, December 13, 2016.
27. Amanda Wood et al., "Reframing the Local–Global Food Systems Debate through a Resilience Lens," *Nature Food* 4 (2023): 22–29.
28. Beale, "Salient Features of the Demography of American Agriculture."
29. Dania Francis et al., "Black Land Loss: 1920–1997," *AEA Papers and Proceedings* 112 (May 2022): 38–42.
30. Bruce McWilliams, "Cost-Benefit Analysis for the Heirs' Property Relending Program Final Rule (FSA-2021-0002-0001)," *U.S. Department of Agriculture, Farm Service Agency*, 2021.
31. MacDonald et al., "Three Decades of Consolidation in U.S. Agriculture"; According to Whitt et al., "America's Farms and Ranches at a Glance," large-scale nonfamily farms accounted for 93 percent of all nonfamily farms' production. They include partnerships of unrelated partners, nonfamily corporations, and farms with a hired manager unrelated to the owners. The value of their production keeps increasing—from 13 percent in 2020 to 17 percent in 2021.
32. James M. MacDonald, "Tracking the Consolidation of US Agriculture," *Applied Economic Perspectives and Policy* 42, no. 3 (2020): 361–79.
33. Whitt et al., "America's Farms and Ranches at a Glance."
34. Lyson et al., "Scale of Agricultural Production," 311–27; Lobao, *Locality and Inequality*.
35. Beale, "Salient Features of the Demography of American Agriculture"; Kirschenmann et al., "Why Worry about the Agriculture of the Middle?"
36. Kenneth M. Johnson and Daniel T. Lichter, "Rural Depopulation: Growth and Decline Processes over the Past Century," *Rural Sociology* 84, no. 1 (2019): 3–27.
37. USDA Economic Research Service, "USDA ERS – Food Security in the U.S."; USDA Economic Research Service, "USDA ERS – Food Security and Nutrition Assistance."
38. USDA Food and Nutrition Service, "Barriers That Constrain the Adequacy of Supplemental Nutrition Assistance Program (SNAP) Allotments (Summary)."
39. Arohi Pathak and Rose Khattar, "Fighting Hunger: How Congress Should Combat Food Insecurity Among Low-Income Americans," *Center for American Progress*, 2022.
40. Desilver, "What's on Your Table? How America's Diet Has Changed over the Decades."
41. Giulia Menichetti et al., "Machine Learning Prediction of the Degree of Food Processing," *Nature Communications* 14 (2023): 2312.
42. Filippa Juul et al., "Ultra-Processed Food Consumption among US Adults from 2001 to 2018," *The American Journal of Clinical Nutrition* 115, no. 1 (2022): 211–21.
43. Centers of Disease Control based on the National Health and Nutrition Examination Survey.
44. Rao et al., "Do Healthier Foods and Diet Patterns Cost More than Less Healthy Options?"
45. Magdoff and van Es, *Building Soils for Better Crops*.
46. Ibid; JoAnn Burkholder et al., "Impacts of Waste from Concentrated Animal Feeding Operations on Water Quality," *Environmental Health Perspectives* 115, no. 2 (February 2007).
47. Magdoff and van Es, *Building Soils for Better Crops*.

48. Ann A. Sorensen and Mitch Hunter, *Wildlife on the Working Landscape: Charting a Way for Biodiversity and Agricultural Production to Thrive Together* (American Farmland Trust, 2020).
49. United States Environmental Protection Agency, "Agriculture," in *Inventory of U.S. Greenhouse Gas Emissions and Sinks: 1990–2020* (2022), 5–60.
50. The Environmental Working Group Farm Subsidy Database tracks farm subsidies from commodity, crop insurance, disaster programs, and conservation payments.
51. Ann Sorensen, *Farms Under Threat*; Daniel Hellerstein et al., "Farmland Protection: The Role of Public Preferences for Rural Amenities," Agricultural Economic Report No. 815. *USDA Economic Research Service*, October 2002.
52. USDA, *Summary Report: 2017 National Resources Inventory* (Washington, DC: Natural Resources Conservation Service and Ames, IA: Center for Survey Statistics and Methodology, Iowa State University, 2020).
53. Julia Freedgood et al., *Farms Under Threat: The State of the States* (Northampton, MA: American Farmland Trust, 2020).
54. C. Barrington-Leigh and A. Millard-Ball, "A Century of Sprawl in the U.S.," *PNAS* 112, no. 27 (2015): 8244–49.
55. Freedgood et al., *Farms Under Threat*.
56. USDA, Economic Research Service. *Farmland Values, Land Ownership, and Returns to Farmland, 2000–2016* (Washington, DC: USDA-ERS, February 2018). 33 pp.
57. Department of Energy, "Wind Vision Detailed Roadmap Actions: 2017 Edition." May 2018.
58. M. Inman, *Planting Wind Energy on Farms May Help Crops, Say Researchers* (National Geographic, December 21, 2011).
59. Franklin D. Roosevelt, Letter to all State Governors on a Uniform Soil Conservation Law. Online by Gerhard Peters and John T. Woolley, The American Presidency Project.
60. This explanation is based on information from a presentation made by AFT Soil Health Program Manager Caro Roszell, September 19, 2022.
61. Ann Sorensen, "Farms Under Threat: The State of America's Farmland."
62. Steve Martinez et al., "Local Food Systems: Concepts, Impacts, and Issues," *USDA Economic Research Service*, 2010.
63. Andrew Dumont et al., "Harvesting Opportunity: The Power of Regional Food System Investments to Transform Communities," *Federal Reserve Bank of St. Louis*, (2017).

2

THE PUBLIC FRAMEWORK FOR FOOD SYSTEMS PLANNING

Chapter Summary

The U.S. Constitution spells out the roles of and relationship between federal, state, and tribal governments. Within this framework, states hold planning authority but generally assign it to local governments. Local governance varies widely from state to state, forming a highly complex web of counties, municipalities, townships, and special districts with varying degrees of autonomy. Further, the federal government recognizes 574 tribes as sovereign, self-governing nations with their own laws and land use authority. Citizens of these tribes have dual citizenship in the states where they live.

This chapter provides an overview of the many tiers of government that affect food systems planning. It introduces common types of planning organizations and describes various types of plans that can play a role in food systems planning. It ends with a discussion of the need to transcend jurisdictional structures to plan for food systems at a regional—or foodshed—scale.

***

Before embarking on a planning process, it is important to understand the governing dynamics of the place—or places—the plan will address. These dynamics affect the scale and scope of planning efforts and the types of actions a plan can suggest. For instance, governance in urban states generally is more complex and regulatory than in rural states with small populations dispersed across large geographies.

In the U.S., the role of government is to safeguard the people's rights and freedoms in a democratic republic where citizens elect representatives to direct public actions. This is not as straightforward as it seems; different people and interest groups interpret rights and freedoms in different ways. Effective planning recognizes and responds to different interpretations within the framework of public authority spelled out in the Constitution.

Article I Section 8 lists the federal government's powers, including the power to declare war, maintain armed forces, coin money, and establish a post office. It also gives it power to regulate commerce with foreign nations, states, and Indigenous tribes. The 10th Amendment gives states all powers not expressly granted to the federal government, including planning authority. States generally share these powers with local governments. Some states are "Home Rule," which gives local governments wide authority and limits the state's power to intervene in local affairs. Others are "Dillon's Rule," a narrower, state-centric approach. Many are both, allowing local governments various degrees of autonomy.

Further, the nation's 333 million people are unevenly distributed across the states in terms of total population size and racial, ethnic, cultural, and socioeconomic characteristics. The largest racial and ethnic group is "White alone non-Hispanic" (White) at 58 percent—down from 65 percent in 2010 as the country becomes more diverse.[1] As demographic changes continue to increase the cultural diversity of communities across the country, it is important to get to know the whole population and address any historical impacts of discrimination that continue to affect food production and community food security.

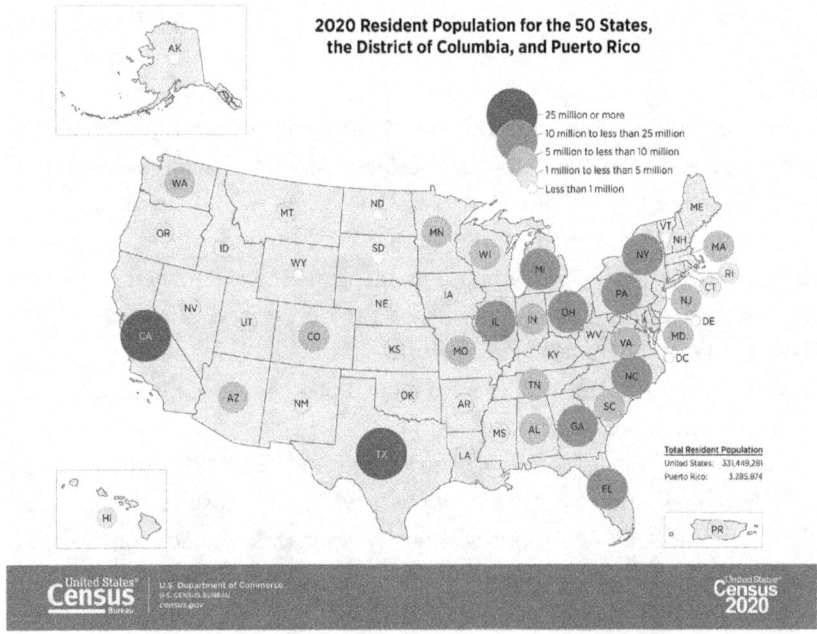

2020 Resident Population for the 50 States Census map

Figure 2.1 The U.S. population is spread unevenly across the states. Some states have large and diverse populations of over 25 million, while others have fewer than 1 million people.

States

States have their own constitutions and substantial independence to adopt laws, policies, regulations, and to determine the form and functions of local government. Thus, local governance varies widely from state to state, forming a highly complex web of counties, municipalities, townships, and special districts. Regional entities also play some governmental roles, including planning, as do sovereign tribes, which have their own planning authorities.

States often share responsibility with local governments for functions like public health or maintaining roads. For example, the interstate highway system is administered by the federal government through the Department of Transportation, but most states have their own departments to oversee intrastate transportation, and local governments maintain local roads. They also both engage in planning activities and delegate it to local governments. Further, to qualify for federal assistance, most states are required to have transportation and hazard mitigation plans.

States also have plans for specific sectors or cross-cutting issues, such as economic and/or workforce development. Twenty-two have climate action plans. Thirteen have adopted strategic or growth management plans with various degrees of oversight and shared authority. A growing number have food plans, although many of these were developed by partnerships and universities instead of government agencies.[2] Very few states have plans for agriculture, although some, like Oregon and Washington, address agricultural land in their state's growth management plan. However, when it comes to food and agriculture policy, states play a larger role. Typically working in the nexus between federal and state priorities, they develop their own policies and administer, inform, and leverage federal programs like the Supplemental Nutrition Assistance Program (SNAP).

The composition of state legislatures and their legislative processes vary. Some have fewer than 100 members, while New Hampshire has more than 400. Most meet annually, but Montana, Nevada, North Dakota, and Texas meet every other year. Ten have full-time legislatures, the rest are part-time. Sessions usually start in January but meet for different lengths of time. The National Conference of State Legislatures website publishes information on the number of legislators and their length of term in every state.[3]

Except for Nebraska, state legislatures have two chambers similar to the federal House and Senate, sometimes called other things. Typically, when bills are introduced, they are assigned to a committee that holds hearings and decides which bills go to the full chamber. If approved, the bill moves to the other chamber, which repeats the process. If the second chamber amends the bill, it goes to a conference committee to resolve differences. Once approved by both chambers, it goes to the governor. If signed, it becomes law; if vetoed the legislature may either sustain or override it.

Tribes

Indigenous tribes are sovereign, self-governing nations that predate colonization and operate independently from federal, state, or local governments. The U.S. officially recognizes 574 tribes, each with their own laws, taxing authorities, and land use authority. Many have constitutions and separation of powers similar to the U.S. and engage in land use and other types of planning.

The Constitution authorizes the U.S. president to make treaties and for Congress to manage relations and regulate commerce with tribes. Some states have compacts and intergovernmental agreements to address issues of shared concern and/or have recognition processes related to tribal status. Similar to state and local governments, tribes receive—or are supposed to receive—federal program assistance. However, their access to federal programs often is impeded by outdated laws, lack of cultural understanding, and bureaucratic red tape. Thus, they often do not receive equal treatment or benefit from these programs.

While states have no authority over tribal governments, citizens of federally recognized tribes have dual citizenship in the states where they live. Subject to the tribal government, they also are subject to the U.S. federal government and eligible for federal program assistance.

Local Governments

Local self-governance is a cherished American tradition, but in reality, local governments have various degrees of autonomy. With powers granted by their state constitutions, this also is true of planning authority. Some states are "Home Rule," which gives local governments wide authority and limits the state's power to intervene in local affairs. Others are "Dillon's Rule," which is a much narrower approach where local governments may only employ powers expressly granted by the state or essential to their own existence. Most states apply both rules in some way.

Some states are mostly Dillon's Rule with limited exceptions, while others are more flexible. Florida provides optional Home Rule, leaving it up to voters to decide whether or not to adopt a county charter. Nebraska and Texas are strict Dillon's Rule states but allow cities with a population of more than 5,000 to form Home Rule governments. In Arizona, Home Rule only applies to cities with at least 3,500 residents (see Figure 2.2).

Counties

Except Connecticut and Rhode Island, all states have some form of county government. In Louisiana, they are called parishes. In Alaska, county equivalents include boroughs, municipalities, and census areas. Counties were created to help states oversee and implement laws, property assessment, and record keeping. As their populations grew, states gave them more authority. Today, 3,143 counties and county equivalents are largely independent. In

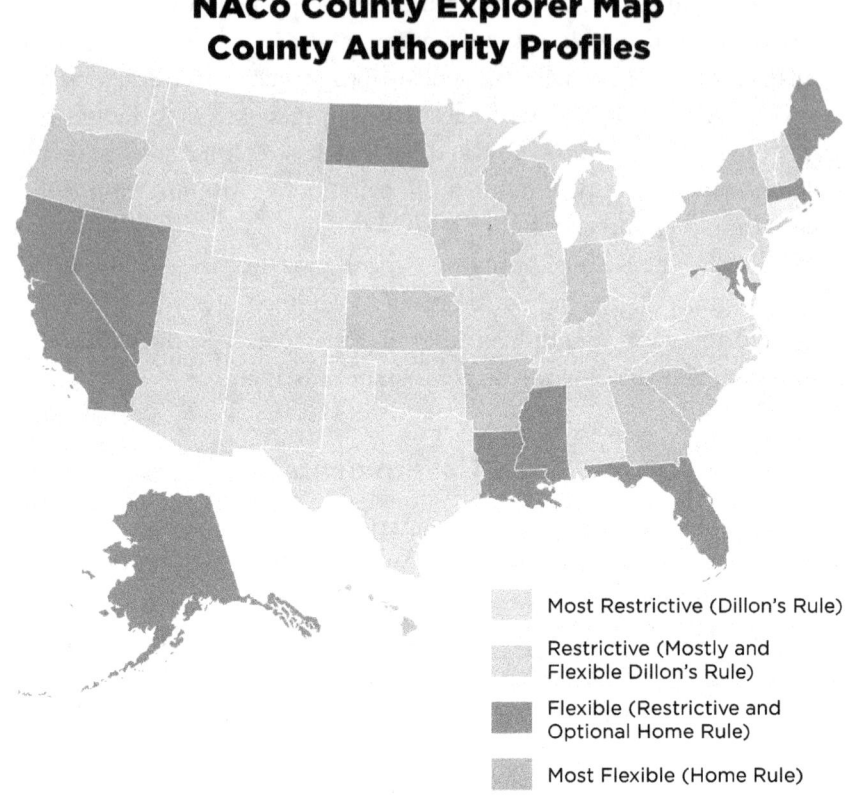

Figure 2.2 This map provides an overview of Dillon's Rule and Home Rule authorities. It is a simplified version of the National Association of Counties (NACo) County Explorer map, a product of a long-term NACO project that provides a more detailed view, is updated over time, and is available in an interactive format with links to state/county authority fact sheets on the NACo website.

most states, they serve as a primary legal division, which gives them policy authority.[4] Most are governed by a popularly elected board such as county commissioners or supervisors. They usually have both executive and legislative authorities, typically related to budget, taxes, and land use policy. Most offer a range of services, from transportation and infrastructure to health care and public safety. Counties also play a statistical role in government and other data collection.

Table 2.1 Five Smallest U.S. Counties by Population Compared to Surface Area

County	State	Population	Surface Area
Kalawao	Hawaii	88	53 square miles
Loving	Texas	113	677 square miles
King	Texas	289	913 square miles
Kenedy	Texas	404	1,946 square miles
Arthur	Nebraska	469	718 square miles

Source: U.S. Census Bureau. Population based on 2020 Decennial Census.

Table 2.2 Five Largest U.S. Counties by Population Compared to Surface Area

County	State	Population	Surface Area
Los Angeles	California	10,137,915	4,751 square miles
Cook	Illinois	5,203,499	1,635 square miles
Harris	Texas	4,589,928	1,778 square miles
Maricopa	Arizona	4,242,997	9,224 square miles
San Diego	California	3,317,749	4,207 square miles

Source: U.S. Census Bureau. Population based on 2020 Decennial Census.

Comparing counties is tricky. They vary widely in size—both in population and land area (see Tables 2.1 and 2.2.) In four states and Puerto Rico, one or more municipalities are independent of county organization but are considered "independent cities" for statistical purposes. So the Census Bureau treats them—and the District of Columbia—as county equivalents.[5]

Municipalities

States grant municipalities authority to incorporate, provide public services, levy taxes, and regulate activities. Geographically within counties, sometimes they cross county lines. Out of nearly 20,000 municipal governments, most have populations of fewer than 10,000, although 10 have more than 1 million residents.[6]

The mayor–council form of governance is the oldest and most common. It separates power between the mayor, who serves as the executive, and the council, which serves as a legislative branch. Mostly in the south and west, an increasing number of municipalities have adopted a council–manager governance structure. Here, an elected council serves as the legislative body to make policy but appoints a professional manager to administer their policies, programs, and services.

Towns and Townships

Towns and townships are both geographic and political subdivisions of the counties in which they are formed. The oldest continuous form of government, they operate in 20 states in three regions: New England, Mid-Atlantic, and Midwest (see Table 2.3).

States grant towns and townships legal authority for governance, which has led to a variety of definitions as well as forms and functions. Most are unincorporated and serve rural areas, often containing multiple towns,

Table 2.3 States With Town and Township Governments

Region	States
New England	• Connecticut • Maine • Massachusetts • New Hampshire • Rhode Island • Vermont
Mid-Atlantic	• New Jersey • New York • Pennsylvania
Midwest	• Illinois • Indiana • Kansas • Michigan • Minnesota • Missouri • Nebraska • North Dakota • Ohio • South Dakota • Wisconsin

Source: U.S. Census Bureau Governments Integrated Directory. See https://www2.census.gov/govs/cog/2007/techdocgovorg.pdf

boroughs, and/or villages. But in New England, towns are the basic unit of local government, similar to townships in other states, but incorporated so they possess municipal powers. In Minnesota, the terms "town" and "township" are used interchangeably. In Iowa, townships have no governance powers of their own. Within this mix of forms and functions, most towns and townships are governed by a small, elected board of supervisors, selectmen, or trustees and oversee local functions such as planning and zoning, administering property taxes and elections, public safety/emergency services, road maintenance, and so on.

Special Districts

Special districts are authorized by state law. Serving as ad hoc entities to provide a limited set of public functions, they have sufficient autonomy to qualify as an independent governmental unit. They can raise revenues through taxes, fees, grants, and/or debt financing. Also called authorities, boards, and commissions, districts typically are created for school, fire, and water services. Also established under state law, *conservation districts* are a little different, as they are formed to carry out local natural resource management programs and to help cooperating land owners and managers protect land and water resources.

Planning Bodies, Partners, and Plans

Planning Bodies

Planning bodies are government authorities responsible for public-sector planning. They range from small, sometimes volunteer boards and commissions to formal planning agencies designated by governors and qualified to receive federal funds. Common entities include the following:

Metropolitan Planning Organizations (MPOs) are formal planning agencies required for cities with populations of 50,000 or more to receive federal transportation funds. They also address land use issues, energy, economic development, environment, and quality of life. They have professional planning staff who also provide policy coordination between state and local governments. MPOs are designated by the governor and form an agreement between state and local governments. Larger urbanized areas with over 200,000 residents are called *Transportation Management Areas* (TMAs). MPOs in TMAs have more sway in setting priorities and are responsible for additional planning products.

Planning commissions and boards typically are advisory bodies that provide citizen review on local planning issues, from comprehensive plans and land use regulations to reviewing development proposals. Some have full-time employees, but many are served by part-time staff and volunteers.

Planning departments. Some states, many counties, and most cities have professional planning departments. At the state level, they often are part of the governor's office or within a department like administration or community/economic affairs, although some have independent planning departments. Most cities have planning departments to develop land use and other plans and programs, and to administer zoning. So do many counties, although their roles and responsibilities differ. In a strong Home Rule state, counties may have planning authority but cities and towns hold authority for zoning.

Regional Planning Commissions/Councils of Government cover multiple jurisdictions to deliver planning functions, coordination, technical assistance, and administer a variety of federal, state, and local programs. They include *commissions, councils of government* (COGs), *regional planning organizations* (RPOs), and *regional purpose authorities* (RPAs). They help facilitate cooperation among local governments, generally related to infrastructure, public health, transportation, and increasingly emergency management.

Regional Transportation Planning Organizations (RTPOs) incorporate rural transportation needs into statewide planning processes. They are formed for multiple jurisdictions in nonmetropolitan areas that have populations of less than 50,000 to identify rural transportation needs and incorporate them into the statewide planning process.

Cooperative Extension

The Cooperative Extension System (CES) is a national education system that operates through state land-grant universities in partnership with federal, state, and local governments. Authorized in 1914 by the *Smith-Lever Act*, its major program areas include agricultural production and business development, community development, emergency preparedness, family and consumer sciences, health and nutrition, and natural resources. While specific programs and priorities differ from state to state, CES supports communities as well as farmers, ranchers, families, and individuals.

CES is supported by state and local governments and shares evidence-based research from state land-grant universities. Extension educators (or agents) provide technical assistance to farmers, ranchers, and communities; develop partnerships across the food supply chain; and provide support in the event of disasters. All these resources combined give CES a vital role in food systems planning and development activities.

The 1862 *Morrill Act* created the land-grant university system, and the 1887 *Hatch Act* directed federal funds to agricultural research. The *Morrill Act of 1890* expanded the system to include historically Black colleges and universities (HBCUs) with programs in agricultural sciences. In 1994, land-grant status was extended to Native American Tribal Colleges through the *Improving America's Schools Act*. The 2008 Farm Bill authorized establishment of a group of Hispanic-serving agricultural colleges and universities to be eligible for USDA Integrated Research, Education, and Extension Competitive Grants Programs.

Common Types of Plans

Sometimes states and communities create stand-alone plans for agriculture and/or food systems. More often, food and farming issues are infused into other planning efforts. What follows are descriptions of types of plans relevant to food systems efforts.

Land Use Plans

Land use plans inform how land use decisions are made. They promote orderly development to meet residents' needs and are most often implemented through zoning and subdivision regulations. Typically, they address six main land use categories: agricultural, residential, recreational, commercial, industrial, and transportation.

Two state enabling acts from the 1920s laid the foundation for land-use planning and zoning. The U.S. Commerce Department published a Standard State Zoning Enabling Act (SZEA) in 1926 and a Standard City Planning Enabling Act (SCPEA) in 1928.[7] The SZEA authorized planning commissions, offered guidance on developing master plans, and provided for approval of public improvements and controls over private land subdivisions. The SCPEA included a provision to divide a local government's territory into districts, along with a statement of purpose and procedures for establishing and amending zoning regulations.

Oregon has the strongest state land use planning law. It requires local governments, special districts, and state agencies to adopt comprehensive plans consistent with 19 Statewide Land Use Planning Goals. States like Arizona require counties to form a planning and zoning commission and to adopt comprehensive plans but do not require state oversight and consistency. Other states, like Georgia and Vermont, do not mandate comprehensive plans but offer incentives through grants and loans.

Comprehensive or Master Plans

Local land use plans are commonly called comprehensive (comp), general, or master plans, and guide land development and protection at the county or municipal level. They have the legal authority to guide community development and a history of practice that builds public acceptance of their policies. The American Planning Association (APA) offers standards and a set of principles and practices to support their preparation.

Comp plans present a vision for the future and lay the foundation for local land use policies. Typically covering from five to 20 years, they reflect the big picture of current conditions and future directions and serve as road maps to guide local decisions that are enforced by policies, incentives, and land use regulations. Some states require them and others do not. Of those that do, some also require that they be updated on a regular basis and/or require which elements to include.

While state requirements vary, most comp plans address major land use categories, including commercial, industrial, residential, and open space. They generally include several key elements: existing conditions, short- and long-term goals, maps of present and future land uses, and land use policies and other implementation strategies. Sometimes they are established as *unified development code*, where the plan and the zoning components are included together as a single document. While most do not address food, some include agriculture or at least farmland, although rarely to the degree they address other land uses.

Agriculture Plans

Dozens of state and local governments have developed stand-alone plans for agriculture, sometimes simply for farmland protection. These plans typically have a land use element to steer growth away from quality soils and active farming areas and establish a framework to protect and conserve farmland and to support agricultural economic development. Maryland's 2019 strategic plan for agriculture was spearheaded by the Maryland Department of Agriculture (MDA) and included an overview of the state's farming sectors, trends, forces, strengths, and challenges affecting them, and a set of initiative areas with recommended actions. MDA's top priority was to ensure an inclusive process that represented all sectors and stakeholders. Together they put forth a vision of high-quality food production, workforce development, consumers who understand agriculture, and support for a new generation of farmers to continue Maryland's legacy as a leader in sustainable farming practices.[8]

A few states have created programs to support local planning for agriculture. New York offers municipal planning grants to protect farmland and county planning grants to promote agricultural viability. County planning grants also may be used to update local planning documents, including but not limited to the agricultural section of comprehensive plans, land use regulations, and zoning ordinances to ensure that these documents contain clear language and policies to support agriculture.[9] Virginia has a planning grant program to support agriculture and forestry by funding local efforts and encouraging local governments and the ag/forestry community to work together to integrate these industries into local economic development.[10] Taking a different approach, California offers Conservation Agriculture Planning Grants to help farmers and ranchers identify actions for climate change mitigation and adaptation, further environmental stewardship, and ensure food security.[11]

PLANNING FOR AGRICULTURE AND FARMLAND PROTECTION IN ERIE COUNTY

Figure 2.3 Erie County's plan for agriculture includes an analysis of trends and conditions, a series of maps, and information about the tools the county can use to support farms and protect farmland from development.
Source: Kevin Keenan.

Erie is a fast-growing county located along the shores of the Great Lake Erie in western New York. It is part of the Buffalo–Niagara Falls metropolitan area, the state's second largest metro area. Its innovative Agricultural and Farmland Protection Plan includes strategies to identify and protect farmland, support new types of farms and attract new farmers, connect farmers with consumers and new markets, and increase access to healthy, local food. Each of its goals is accompanied by specific implementation actions backed up by an analysis of current conditions and trends, a series of maps, and information about ways to support farms and protect farmland.

Erie's plan for agriculture resulted from a nearly two-year, collaborative process that included eight public meetings to facilitate community input on goals, strategies, and recommended actions. Committed to transparency, results were shared on the county website and in a series of summaries included as part of the plan, along with related planning documents and a set of appendices.[12]

Climate Action Plans

States have begun to adopt climate action plans (CAPs) to mitigate GHG emissions and adapt to the impacts of a changing climate. CAPs are strategic documents that include recommendations for policies and actions. The Center for Climate and Energy Solutions publishes a map that includes a brief description of and links to these plans.[13]

Since the causes and impacts of climate change are intertwined with land use and transportation, plans generally take a holistic approach to help communities prepare for resiliency. California and Massachusetts serve as good examples. California's 2022 Climate Change Scoping Plan lays out an aggressive, sector-by-sector roadmap to help the state achieve carbon neutrality by 2045. Key strategies include updating the cap-and-trade program to align with reduced GHG emission goals, phasing down oil and gas operations, and conserving 30 percent of the state's natural and working lands to capture and store emissions while supporting healthy ecosystems.[14] Massachusetts' Municipal Vulnerability Preparedness (MVP) program provides support for cities and towns to identify climate hazards, assess vulnerabilities, and develop action plans to improve their resiliency. Communities that complete the program become certified and eligible for MVP Action Grant funding and other opportunities.[15]

PLANNING FOR REGIONAL RESILIENCE IN FLORIDA

The Tampa Bay Regional Resilience Action Plan addresses food systems and other community issues in the context of climate change. Covering seven Florida counties, it is an action plan for coordinated preparation and adaptation to impacts such as sea level rise, severe hurricanes, extreme heat, and other weather-related hazards. Within a context of addressing complex land use issues, it strives for more sustainable and resilient food systems and improved access to healthy foods. Toward these ends, it includes strategies such as developing an inventory of agriculture lands, number of farms, vacant lots and production outputs, developing incentives to increase local food production, processing and distribution, and implementing outreach and education efforts to reduce food waste.[16]

Emergency Management/Hazard Mitigation Plans

Hazard mitigation plans minimize the impact of natural disasters by identifying risks and vulnerabilities and developing long-term strategies to protect people and property. Underpinned by federal legislation, they are the foundation for plans that integrate resilience and long-term risk reduction.

The Disaster Mitigation Act of 2000 (which amended the 1988 Stafford Act)[17] creates a framework for state, local, tribal, and territorial governments to receive federal assistance for hazard mitigation planning. The Federal Emergency Management Agency (FEMA) provides guidance to support its requirement that policies are reviewed and revised on a regular basis. Its *State Mitigation Planning Policy Guide* provides state hazard mitigation planning requirements. Beyond FEMA, disaster planning and policy are supported by the Department of Homeland Security (DHS) and the Occupational Safety and Health Administration (OSHA). DHS publishes guidance for public, animal, and plant health agencies who respond to food and agricultural incidents at all levels of government. OSHA gives information to help farmers and others prepare emergency action plans to respond to workplace emergencies and disasters.

When it comes to food, most emergency planning is directed at individual households, not communities. However, the Food and Agriculture Defense Initiative: Extension Disaster Education Network (EDEN) provides national coordination for outreach to enhance the biosecurity of U.S. agriculture and food systems. A collaborative multistate Cooperative Extension effort, EDEN's mission is to reduce the impact of disasters through research-based education.

In response to the food supply breakdowns of Covid-19, Ohio State University researchers established partnerships with several state agencies to study the relationship between food systems and emergency management. They found many policies and programs administered by different agencies and organizations with little or no connection to each other. Their report, "Preparing for Food System Resiliency in Ohio: Policy and Planning Lessons from COVID-19" includes key takeaways for short-, medium-, and long-term food system resilience planning.[18] Engaging EDEN into food systems planning could open opportunities to learn from and build on research and extension efforts and include food systems in emergency management plans.

> **A TEMPLATE FOR EMERGENCY FOOD PREPAREDNESS**
>
> The Alaska Food Policy Council created a Community Food Emergency and Resilience Template to help communities plan for emergency events. Providing a framework to design a community emergency food preparedness response plan and strategy, it outlines procedures and resources to address food security in the event of natural disasters, supply chain disruptions, production failures, and other emergencies. The template serves as a model to help communities identify transportation requirements; storage capability; food reserves; and procedures to release, obtain, and provide food for community residents and addresses local production, hunting, gathering, and fishing.[19]

Food Systems Plans

Food systems plans address a set of interconnected, forward-thinking activities to strengthen local and regional food systems through policies, procedures, and public investments. "Eating Here: Greater Philadelphia's Food System Plan" was one of the first. Released in 2011 by the Delaware Valley Regional Planning Commission, it covered nine counties in Pennsylvania and New Jersey addressing concerns from securing a supply of farmland to grow food to the nutrition and health of the consumers who eat it.

Vermont's Farm to Plate (F2P) is a comprehensive state-level plan. Led by the Vermont Jobs Fund, its goals are to increase economic development and employment in the farm and food sectors, improve the resilience of the working landscape, and increase access to healthy local foods. Updated in 2019, F2P works to achieve a 2030 vision for Vermont's food system through 15 strategic goals, 87 objectives, and 34 recommended actions. It also includes a series of briefs highlighting current conditions, bottlenecks, gaps, opportunities, and recommendations.[20] In a related effort, the New England State Food System Planners Partnership launched a ten-year planning initiative to prepare the six-state region for system shocks, including weather events and public health emergencies. "New England Feeding New England: Cultivating A Reliable Food Supply" lifts up food systems plans developed by each of the states to create a regional strategy.

Municipalities also have created food system plans—although most are not comprehensive, focused more on food access than food systems.[21] One exception is the City and County of Denver, which developed a strategic *Denver Food Vision* detailing implementation areas with specific strategies to achieve 12 measurable goals by 2030.[22] Further, cities like Baltimore and Boston also have created *food systems resilience plans* (see box).

Food systems plans often are driven by nonprofit organizations. For example, Arizona's food policy council released a three-year statewide food action plan to inform advocacy and build capacity and investment to transform Arizona's food system. The plan addresses four priorities: food access and distribution, land and water access and protection, climate smart foodways, and workforce development. Each has a goal with objectives, strategies, and actions to achieve it.[23] In Niagara Falls, New York, a Healthy Food Healthy People working group spearheaded a resident-based planning process to create a Local Food Action Plan that ultimately was adopted by the city council. The plan includes four priority areas, including agriculture, economic development, education, and healthy neighborhoods, each supported by sub-themes and action steps to improve food access and security.[24]

RESILIENT FOOD SYSTEMS, RESILIENT CITIES

Boston, Massachusetts's "Resilient Food Systems, Resilient Cities" plan was developed after two destructive weather events: Hurricane Sandy in 2012 and record-breaking snowfall in 2015. Calling for strong public–private partnerships and coordination across various city agencies, it offers seven overarching recommendations to build a sustainable regional food economy that can withstand weather-related disruptions. Each overarching recommendation includes a range of suggested actions. Examples include:

- Establish a food system resilience committee to strengthen coordination between local food system initiatives and resilience planning;
- Diversify the milk supply to retail outlets and review distributors' contingency plans;
- Encourage the growth of smaller regional suppliers and new processing facilities;
- Develop neighborhood food availability action plans; and

- Commission studies to better understand neighborhood food consumption patterns and revisit the one-mile radius to food retail and emergency food outlets after major storms.

It also has sets of recommendations related to emergency food, food storage, road infrastructure, areas prone to flooding, and improving coordination between city plans and agencies.[25]

Sustainability Plans

Sustainability plans address issues related to environmental management for concerns such as air and water quality, energy, and waste. Typically, they balance environmental and economic conditions with social equity and other community priorities.

APA outlines objectives for sustainability planning to reduce dependence on activities that harm ecosystems. Broader in scope than CPAs, sustainability plans address environmental, economic, and social considerations, including and beyond those directly related to climate change. Los Angeles is a good example of how to integrate food systems into a local sustainability plan.

INTEGRATING FOOD SYSTEMS INTO SUSTAINABILITY PLANNING

"Our County" is a bold, regional sustainability plan for Los Angeles, California. Organized around 12 cross-cutting goals, it outlines what the county of 10 million residents and 88 municipal governments can do to address compound environmental challenges within a complex tapestry of local government authority. Among its goals, it seeks to improve healthy-food access by leveraging its capital assets, public services, and regulatory authority while optimizing its purchasing power and business services to make food production more sustainable.

Stretching along 75 miles of coastline, LA includes two offshore islands, national forests, and the Mojave Desert. So "Our County" emphasizes the importance of its rural communities, ecosystems, habitats, and biodiversity along with climate change. Reckoning with the county's history

> of exclusionary zoning, racial covenants, and siting toxic land uses in communities of color, "Our County" also has economic and equity goals and commits to helping communities affected by environmental pollution and inadequate and unmaintained infrastructure.[26]

Planning for Food Systems at a Regional Scale

Given the structures of governance and planning, it is unsurprising that planning for food and agriculture usually occurs at state or local levels. However, for sustainability and resiliency, taking a regional—or foodshed—approach can return lasting benefits. Drawn from the concept of watersheds, foodsheds are metaphors for geographic regions that link agriculture and food production to the dietary needs of people in nearby population centers.[27] Like watersheds, they transcend jurisdictional boundaries.

Operating at a foodshed scale expands production capacity by increasing the diversity of soil, water, microclimate, and other natural resource factors. Containing different topographies and scales, a foodshed can serve large populations and offer an abundant, varied, and culturally appropriate food supply. Taking a regional approach expands access to supply chains and markets, leverages urban–rural linkages, reduces the need for imports, and keeps and more equitably distributes economic returns within both rural and urban communities.[28] While it can be hard to overcome territorial barriers, it is a good time to address the challenge and facilitate dialogue to address potential tensions and move forward on opportunities.

During the pandemic, nimble local and regional networks and flexible supply chains protected communities from food shortages. Still, rigidity in the current policy and regulatory environment constrained a more strategic response.[29] These weaknesses can be addressed through foodshed-scale planning.

Regional approaches also can help reconcile cultural differences between urban and rural communities and their attitudes toward planning, policy, and the role of government. Most rural residents believe they have different values from people living in cities,[30] and urban planning approaches often do not translate well to rural communities. To date, most food systems planning has focused on urban areas and implementation has emphasized zoning and land use regulations for community gardens, city farms, and local procurement.[31] While these all are valuable, they will not ensure food security much less resiliency for urban food supplies, much less food systems generally. More and better planning is needed to expand the middle infrastructure to support rural agriculture, protect farmland for food production, and incentivize food aggregation, processing, and distribution outside of urban centers.

That said, most descriptions of urban agriculture are more expansive than community gardens and inner-city farms. They connect production, processing, and distribution of food and other farm products in urban, suburban, and peri-urban areas with feeding local populations, expanding green spaces, workforce development, and fostering community education and engagement.[32] Adopting this frame for food systems planning would help bridge urban and rural communities, opening the door to more regional or foodshed approaches.

WHAT IS URBAN AGRICULTURE?

Figure: 2.4 Bobby Wilson established Metro Atlanta Urban Farming in 2009 to address the growing need for affordable food in low-income communities. His five-acre farm in College Park, Georgia, is just minutes from downtown Atlanta.
Source: USDA. Photo by Preston Keres.

> Urban agriculture is defined differently in different places and policies, often dictated by program funding. While it has neither a statutory nor formal definition, a seminal definition from 2000 provides a useful frame: "The growing, processing, and distribution of food and other products through intensive plant cultivation and animal husbandry in and around cities."[33] USDA has a similar definition: "The cultivation, processing and distribution of agricultural products in urban and suburban areas. Community gardens, rooftop farms, hydroponic, aeroponic, and aquaponic facilities, and vertical production are all examples of urban agriculture. Tribal communities and small towns may also be included."[34]
>
> Despite these fairly broad definitions, urban agriculture often is limited to community and rooftop gardens rather than intensive food production outside of city centers but within urban-influenced areas. While it may be harder to enact policies that cross municipal and county boundaries, taking a broader approach to urban agriculture—especially in a food systems planning context—will at once increase food security and agricultural viability.

Urban-influenced counties produce most of the perishable foods we eat, including the vast majority of fruits and vegetables, two-thirds of dairy, and more than half of poultry and eggs[35] (see Figure 2.5). These places include some of the nation's most abundant farming regions, like California's San Joaquin Valley. They also supply two-thirds of direct-to-consumer food sales.[36] Some appear more rural than others, but they all possess attributes of rural communities. Engaging these communities in food systems planning is vital to future food security and climate resiliency.

Rural counties provide most of the country's staple foods, water, energy, and outdoor recreation.[37] They produce wheat, rice, barley, and oats, as well as feed for poultry and livestock. They also provide grazing on millions of acres that otherwise would not sustain food production. And agriculture provides nearly 17 percent of employment in highly rural and remote areas.[38]

While rural America is not homogenous, it has similar traits that affect planning practice. Its rich natural resources require land management along with land use. Its stakeholders include farmers and ranchers whose schedules are weather-dependent and do not work from 9 to 5. Rural residents suffer disproportionate rates of poverty, unemployment, and food insecurity.

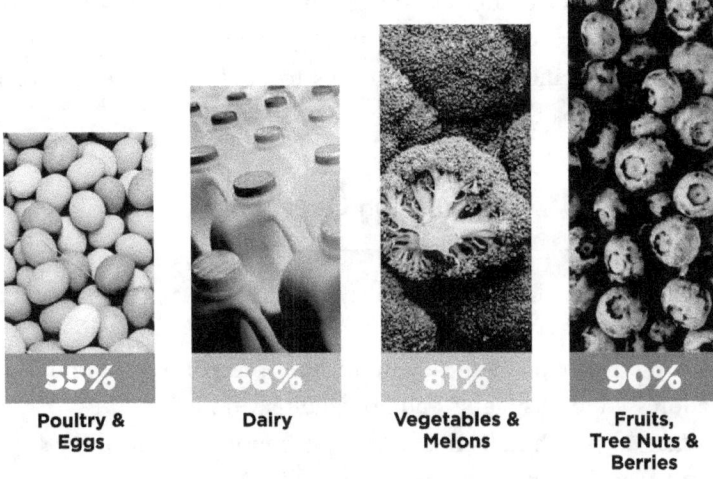

Food Produced in Urban-Influenced Counties
By Percent of Market Value

55%	66%	81%	90%
Poultry & Eggs	Dairy	Vegetables & Melons	Fruits, Tree Nuts & Berries

Source: American Farmland Trust's Farmland Information Center, using data from USDA, National Agricultural Statistics Service, 2017 Census of Agriculture and USDA Economic Research Service 2013 Urban Influence Codes.

Figure 2.5 Much of the food we eat comes from urban-influenced counties. Source: American Farmland Trust's Farmland Information Center.

Of 310 counties with high and persistent levels of poverty, 86 percent are rural. Along with limited access to healthy, affordable food, many face barriers to medical care and internet services. Black and Indigenous populations are especially vulnerable.[39]

More and better rural planning is needed. But rural communities tend to be under-resourced and poorly understood. With small populations and limited government, they often lack planning capacity. What planning is done often is done by volunteer boards and commissions or by outside planning consultants. If they have paid staff, they often have multiple other responsibilities unrelated to planning. Further, rural populations often distrust government intervention, so policy—especially regulatory—changes can be hard to achieve.

Taking a foodshed approach can bring resources to the table that may otherwise be unavailable to communities planning on their own. For rural communities, this could include planning expertise and new market opportunities. For urban communities, this could include a fresher, more reliable source of food. Taking an assets-based approach like the Community Capitals Framework can identify and leverage community resources to build wealth and respond to food system needs. Chapter 3 explains this framework and offers principles and practices to guide food systems planning along the rural–urban continuum.

Discussion Questions

1. What are the planning and policy dynamics in your state/community?
2. What types of plans already in place in your state/community have addressed food systems and how did they do it?
3. What do you think is needed to foster more sustainable and resilient food systems where you work or live?

Notes

1. U.S. Census Bureau, "U.S. Census Bureau QuickFacts: United States," accessed January 29, 2023; Jensen et al., "The Chance That Two People Chosen."
2. Hoey et al., *Participatory State and Regional Food System Plans and Charters*.
3. National Conference of State Legislatures, "Number of Legislators and Length of Terms in Years," April 19, 2021.
4. U.S. Census Bureau, "Metropolitan and Micropolitan Statistical Areas Population Totals and Components of Change: 2020–2021," *Census.gov*, accessed January 29, 2023; GIS Geography, "13 Open Source Remote Sensing Software Packages," *GIS Geography*, March 15, 2023.
5. Naya El Nassar, *More Than Half of U.S. Population in 4.6 Percent of Counties* (Washington, DC: U.S. Census Bureau, October 24, 2017).
6. "Number of U.S. Cities, Towns, Villages by Population Size 2019," *Statista*, accessed March 18, 2023.
7. Department of Commerce Advisory Committee on Zoning, "Standard State Zoning Enabling Act and Standard City Planning Enabling Act," *American Planning Association*, 1928.
8. Maryland Department of Agriculture, "Strategic Plan for Maryland Agriculture," *Maryland Department of Agriculture*, December 1, 2019.
9. For information about New York's planning grants programs for agriculture and farmland protection visit: "Farmland Protection Planning Grants Program," *New York State Department of Agriculture and Markets*, accessed March 18, 2023.

10. For information about Virginia's Agriculture and Forestry Industries Development Fund (AFID) Planning Grant program visit: "AFID Planning Grants," Virginia Department of Agriculture and Consumer Services.
11. For more information about California's Conservation Agriculture Planning Grants Program, visit: "CDFA – OEFI – Conservation Agriculture Planning Grants Program," *California Department of Food and Agriculture*.
12. American Farmland Trust, "Erie County Agricultural and Farmland Protection Plan," *Farmland Information Center*, October 24, 2012.
13. Center for Climate and Energy Solutions, "U.S. State Climate Action Plans," December 2022.
14. California Air Resources Board, "2022 Scoping Plan for Achieving Carbon Neutrality," *California Air Resources Board*, November 16, 2022.
15. For information about Massachusetts' Municipal Vulnerability Preparedness program visit: "Municipal Vulnerability Preparedness (MVP) Program," Mass.Gov.
16. Cara Serra, *Regional Resiliency Action Plan* (Pinellas Park, FL: Tampa Bay Regional Planning Council, November 2022).
17. Formally, the Stafford Act is known as the Robert T. Stafford Disaster Relief and Emergency Assistance Act, PL 100–707, and was signed into law on November 23, 1988, amending the Disaster Relief Act of 1974, PL 93–288. It provides statutory authority for most federal disaster response activities, especially those related to FEMA programs.
18. Shoshanah Inwood et al., *Preparing for Food System Resiliency in Ohio: Policy and Planning Lessons from COVID-19*.
19. Alaska Food Policy Council, "Community Food Emergency and Resilience Template," *University of Alaska Fairbanks Cooperative Extension Service*, September 27, 2015.
20. To learn more about Vermont Farm to Plate, visit: Vermont Sustainable Jobs Fund (VSJF), "Vermont Farm to Plate Strategic Plan."
21. Jane Karetny et al., "Planning toward Sustainable Food Systems: An Exploratory Assessment of Local U.S. Food System Plans," *Journal of Agriculture, Food Systems, and Community Development* 11, no. 4 (September 2, 2022): 115–38.
22. Denver Department of Public Health and Environment, *Denver Food Action Plan* (Denver, CO: Denver Department of Public Health and Environment, June 2018).
23. Arizona Food Systems Network, "Arizona Statewide Food Action Plan 2022–2024," *Arizona Food Systems Network*.
24. Healthy Food Healthy People Work Group, "Niagara Falls Local Food Action Plan," 2017.
25. Kim Zeuli, Austin Nijhuis, and Pete Murphy, *Resilient Food Systems, Resilient Cities: Recommendations for the City of Boston* (Boston, MA: Initiative for a Competitive Inner City, May 2015).
26. Los Angeles County Chief Sustainability Office, "Our County – Los Angeles Countywide Sustainability Plan," *Our County*, August 6, 2019.
27. Julia Freedgood, Marisol Pierce-Quiñonez, and Kenneth Meter, "Emerging Assessment Tools to Inform Food System Planning," *Journal of Agriculture, Food Systems, and Community Development* 2, no. 1 (December 23, 2011): 83–104.
28. Kathryn Z. Ruhf and Kate Clancy, "A Regional Imperative: The Case for Regional Food Systems," *Thomas A. Lyson Center for Civic Agriculture and Food Systems*, September 20, 2022.

29. Thilmany et al., "Local Food Supply Chain Dynamics and Resilience during COVID-19."
30. Jose A. DelReal and Scott Clement, "Rural Divide: New Poll of Rural Americans Shows Deep Cultural Divide with Urban Residents," The Washington Post, June 17, 2017.
31. Clark, Conley, and Raja, "Essential, Fragile, and Invisible Community Food Infrastructure."
32. Leslie Glover, "Office of Urban Agriculture and Innovative Production Overview." USDA, "Urban Agriculture"; Carolyn Dimitri, Lydia Oberholtzer, and Andy Pressman, "Urban Agriculture: Connecting Producers with Consumers," ed. Fabio Verneau and Professor Christopher J. Griffith, British Food Journal 118, no. 3 (January 1, 2016): 603–17; Martin Bailkey and Joe Nasr, "From Brownfields to Greenfields: Producing Food in North American Cities," Community Food Security News, Fall 2000, 6; Anu Rangarajan and Molly Riordan, "The Promise of Urban Agriculture, National Study of Commercial Farming in Urban Areas," USDA Agricultural Marketing Service and Cornell University Small Farms Program, August 2019.
33. Bailkey and Nasr, "From Brownfields to Greenfields."
34. USDA, Topics, "Urban Agriculture."
35. Freedgood et al., Farms Under Threat, 25–30.
36. USDA National Agricultural Statistics Service, 2012 Census of Agriculture Highlights: Direct Farm Sales of Food (Washington, DC: USDA National Agricultural Statistics Service, December 2016).
37. U.S. Census Bureau and America Counts Staff, "What Is Rural America? One in Five Americans Live in Rural Areas," Census.gov, August 9, 2017.
38. John Newton, "USDA's Early Look at 2019 Farm Income: USDA's Next Update Set for August 2019," American Farm Bureau Federation, March 6, 2019.
39. Tracey Farrigan, Rural Poverty & Well-Being (Washington, DC: USDA Economic Research Service, November 29, 2022).

3

PRINCIPLES AND PRACTICES TO GUIDE PLANNING FOR FOOD AND AGRICULTURE

Chapter Summary

Planning is highly decentralized and occurs at many levels of government. Building on lessons learned from the various types of plans described in Chapter 2, this chapter offers basic principles and practices to guide planning for farms and food, whether in stand-alone plans, or as elements of other types of plans.

***

As discussed in Chapter 2, states have planning authority but generally delegate it to local governments. Food system issues may be infused into many different types of plans—from comprehensive plans to stand-alone plans for food and/or agriculture. Many types of entities address the people, processes, and policies responsible for producing, packaging, processing, distributing, acquiring, consuming, and disposing of foods and food products. They range from official planning departments, boards, and commissions to "small p" planners from nonprofit organizations, Extension, food policy councils, and the like. Working with or within the government, planners address how various public priorities work together and create plans to achieve them through policies, programs, and public investments. This involves imagining what can happen in the next three to five years, or as far out as 20 or more years into the future.

Guiding Principles

Principles drive practices: Who to include in the planning process, how to engage them, and what questions to ask, data to analyze, and recommendations to make. Multifaceted and complex, food systems cut across multiple sectors. Thus, planning for food systems benefits from a *systems*—not individual sectors—approach.

Several organizations have developed principles to guide food systems planning. The Academy of Nutrition and Dietetics, American Nurses Association, American Planning Association, and American Public Health Association drafted seven shared principles to support socially, economically, and ecologically sustainable food systems. They range from supporting the physical and mental health of all farmers, workers, and eaters to accounting for public health impacts across the entire life cycle of food. They also include conserving, protecting, and regenerating natural resources, landscapes, and biodiversity.[1] The Global Alliance for the Future of Food advances similar principles: renewability, resilience, health, equity, diversity, inclusion, and interconnectedness.[2] These principles also reflect those in the Code of Ethics and Professional Conduct for certified planners.

Together, these point to a shared set of guiding principles:

- Diversity in size, scale, geography, practices, preferences, and culture;
- Resilience in the face of challenges and disruptions;
- Individual, community, and ecological health;
- Access to food, land, water, and other essential resources; and
- Transparency and fair treatment throughout the supply chain.

Along with these overarching principles are a few practical ones to guide the planning process.

I. Focus on Assets Not Deficits

Taking an asset-based approach is especially important in rural and lower-wealth communities, which typically are defined by what they lack—like population or financial resources. But they may have other valuable assets, like natural resources, close social ties, and cultural heritage. As it can be hard for people to identify their assets, the Southern Rural Development Center has developed a curriculum to help communities map their assets.[3]

Asset-based strategies marshal community resources to create opportunities and respond to needs. They use informal as well as formal research, and engage with local organizations, associations, and networks to identify resources and move plans forward in ways that are stakeholder-centered and community empowering. The Community Capitals Framework is a trusted approach. Created by Cornelia and Jan Flora[4] for use in rural communities, it also is a useful way to identify and harness food systems assets across the rural–urban continuum.

> **THE SEVEN COMMUNITY CAPITALS—OR ASSETS**
>
> 1. Financial capital: Banks, community development corporations and financial institutions, credit unions, funding for loans and venture capital, grants, and other resources that contribute to wealth creation.
> 2. Built capital: Buildings, roads, bridges, and other supporting infrastructure from food hubs and markets to public water and sewage treatment facilities.
> 3. Social capital: *Horizontal linkages* and relationships between people, including *bonding* activities between people who are close to each other through family, work, worship, and other ties, and *bridging* activities between people and groups who are not closely tied. *Vertical linkages* harness resources from outside the community to support local goals and activities.
> 4. Human capital: Education, training, workforce development, and other activities that lead to knowledge and skills, as well as community-based institutions like health care, childcare, and schools that support them.

> 5. Cultural capital: Beliefs, values, symbols, tastes, and preferences reflected in things like attitudes, art, customs, clothing, language, and food as well as things like fairs, festivals, and museums.
> 6. Natural capital: The full range of ecosystem features from land and soil to water and microclimates to wildlife habitat and biodiversity.
> 7. Political capital: Access to power and the ability to influence decision-making.

II. Engage Early, Often, and Throughout the Process

It is often said that change happens at the speed of trust. Trust is built with time, respect, and engagement. Many planning efforts fall short because they rush to extract information from stakeholders instead of involving them in a meaningful way throughout the process.[5]

Food systems planning efforts are strengthened and sustained when guided by a shared vision and values based on a community engagement process that aligns actions with values and celebrates local wisdom and knowledge. The heart of the process is recognizing and including a full range of community interests, or stakeholders, affected by the plan—from farmers, food-insecure populations, and people who represent the middle infrastructure in between.

Many plans rely on outside consultants, but it still is good to have people embedded in the community guide the engagement process. It also is important to be inclusive from the start—not just inviting people to a table you set, but including them in both building and setting the proverbial table. Once key stakeholders are engaged, make sure they feel welcomed, valued, and heard, especially when they have opposing perspectives.

III. Plan With Implementation in Mind

A good plan serves the community and builds support for action. It has little value if it is not implemented.

Community needs, desires, demands, and opportunities vary widely, partly because of governing structures and political attitudes, and also because of environmental, economic, and population characteristics. There is no silver bullet. Implementation must be grounded in the unique mix of strategies and actions that resonate with people in each individual place.

Building off assets, but recognizing constraints, effective food systems plans integrate multiple sectors and interests. Balancing these takes time and effort and includes innovative as well as tried-and-true strategies to improve economic, environmental, and public health outcomes.

Planning Practices

This section is informed by decades of experience working with and learning from planning processes in a wide range of places, both rural and urban, across the U.S. The lessons learned are drawn from the author's experience planning for agriculture in states and communities across the U.S. and from Growing Food Connections and Pathways to Prosperity both multi-year, collaborative research and extension food systems projects supported by USDA's National Institute of Food and Agriculture.

Getting Started

Before embarking on a planning process, it is good to assess the community's readiness for change and decide on a leadership structure. Together, these increase the likelihood that the process will be inclusive and that the plan will result in actions.

Readiness is the extent to which a community is prepared to act on an issue or a plan. It tends to be issue specific: Communities typically are willing to address some issues and take certain kinds of actions but not others. For example, a rural community may be ready to address agricultural viability but not community food security, where an urban community may be the other way around.

Planning processes have various leadership and decision-making structures. From the outset, it is important for leaders to communicate these to the people involved in the process, along with the scope of work, and a realistic timeline. Leadership may come from a planning department, planning board or commission members, outside consultants, Extension, or a nonprofit or other organization and may include representative stakeholders and people with community connections, specific expertise, or skills like Geographic Information System (GIS) mapping.

It also is important to understand the place or places affected by the plan. Food production capacity, population dynamics, and other factors vary widely across the country. With its Mediterranean climate, California is ideally suited for intensive food production. It also is the third-largest state geographically, with the largest and second most diverse population in the country, and fierce development pressure. In contrast, Alaska is by far the largest state geographically. With the third smallest population overall, it has the largest Indigenous population. Most of its land is publicly owned, and its harsh climate and poor soils limit agriculture to a few regions.[6] Other factors to consider include population growth, which is significant in states like Utah and North Dakota, and population decline in states like West Virginia. Table 3.1 highlights some key differences between states to caution against cookie-cutter approaches.

Table 3.1 Population Characteristics, Gross Farm Receipts, and Total Land and Water Area by State

States	Total Population 2020	Population Change Percent	Diversity Index* Percent	Gross Farm Receipts 2021	Total Area Sq Mi
Alabama	5,024,279	5.1	53.1	$6,831,141	52,420
Alaska	733,391	3.3	62.8	$71,191	665,384
Arizona	7,151,502	11.9	61.5	$4,408,666	113,990
Arkansas	3,011,524	3.3	49.8	$11,467,018	53,179
California	39,538,223	6.1	69.7	$54,506,380	163,695
Colorado	5,773,714	14.8	52.3	$9,528,528	104,094
Connecticut	3,605,944	0.9	55.7	$658,765	5,543
Delaware	989,948	10.2	59.6	$1,494,521	2,489
Florida	21,538,187	14.6	64.1	$8,980,922	65,758
Georgia	10,711,908	10.6	64.1	$10,942,842	59,425
Hawaii	1,455,271	7	76	$687,577	10,932
Idaho	1,839,106	17.3	35.9	$9,068,110	83,569
Illinois	12,812,508	-0.1	60.3	$25,218,484	57,914
Indiana	6,785,528	4.7	41.3	$15,496,595	36,420
Iowa	3,190,369	4.7	30.8	$40,550,005	56,273
Kansas	2,937,880	3	45.4	$24,472,285	82,278
Kentucky	4,505,836	3.8	32.8	$7,713,729	40,408
Louisiana	4,657,757	2.7	58.6	$4,187,292	52,378

(Continued)

Table 3.1 (Continued)

States	Total Population 2020	Population Change Percent	Diversity Index* Percent	Gross Farm Receipts 2021	Total Area Sq Mi
Maine	1,362,359	2.6	18.5	$899,573	35,380
Maryland	6,177,224	7	67.3	$2,838,319	12,406
Massachusetts	7,029,917	7.4	51.6	$578,263	10,554
Michigan	10,077,331	2	45.2	$10,521,249	96,714
Minnesota	5,706,494	7.6	40.5	$24,059,926	86,936
Mississippi	2,961,279	-0.2	55.9	$6,806,852	48,432
Missouri	6,154,913	2.8	40.8	$13,818,290	69,707
Montana	1,084,225	9.6	30.1	$4,481,701	147,040
Nebraska	1,961,504	7.4	40.8	$29,633,171	77,348
Nevada	3,104,614	15	68.8	$952,454	110,572
New Hampshire	1,377,529	4.6	23.6	$219,784	9,349
New Jersey	9,288,994	5.7	65.8	$1,492,708	8,723
New Mexico	2,117,522	2.8	63	$3,508,228	121,590
New York	20,201,249	4.2	65.8	$6,549,978	54,555
North Carolina	10,439,388	9.5	57.9	$14,948,590	53,819
North Dakota	779,094	15.8	32.6	$12,142,319	70,698
Ohio	11,799,448	2.3	40.0	$12,930,854	44,826
Oklahoma	3,959,353	5.5	59.5	$8,901,266	69,899
Oregon	4,237,256	10.6	46.1	$6,172,864	98,379
Pennsylvania	13,002,700	2.4	44.0	$8,297,124	46,054
Rhode Island	1,097,379	4.3	49.4	$76,733	1,545
South Carolina	5,118,425	10.7	54.6	$3,153,893	32,020
South Dakota	886,667	8.9	35.6	$14,093,947	77,116
Tennessee	6,910,840	8.9	46.6	$4,765,716	42,144
Texas	29,145,505	15.9	67.0	$29,278,492	268,596
Utah	3,271,616	18.4	40.7	$2,233,343	84,897
Vermont	643,077	2.8	20.2	$887,011	9,616
Virginia	8,631,393	7.9	60.5	$4,412,950	42,775
Washington	7,705,281	14.6	55.9	$11,320,197	71,298

(Continued)

Table 3.1 (Continued)

States	Total Population 2020	Population Change Percent	Diversity Index* Percent	Gross Farm Receipts 2021	Total Area Sq Mi
West Virginia	1,793,716	-3.2	20.2	$787,555	24,230
Wisconsin	5,893,718	3.6	37	$14,506,357	65,496
Wyoming	576,851	2.3	32.4	$1,944,636	97,813

Note: *The Diversity Index refers to the percent chance that two people chosen at random will be of different racial or ethnic groups.

Sources: U.S. Census Bureau 2010, 2020 for population and land area; USDA ERS for farm sector financial indicators, state rankings, 2021.

Planning Process

Planning is a messy, iterative process. Creating an effective planning process starts with reflecting on who needs to be involved and in what ways. It is important to include representatives who will be most affected by the plan and any resulting policies. If it addresses food production, this includes farmers, including but not limited to those involved with mainstream agricultural organizations. If it addresses food security, this includes food access advocates and representative eaters, especially people living in areas with limited access to supermarkets and other sources of nutritious food. It also includes staff from support institutions like Extension, economic development authorities, and lenders, as well as people who represent middle infrastructure interests, like food processors, purchasers, and distributors.

Beyond including people and organizations from your own networks, consider who is connected to them through other networks. They can help you determine and reach out to important interests who may have been left out. It helps to create and maintain a stakeholders database to manage relationships and streamline engagement. Once you have identified key stakeholders, include them in the planning process to ensure you address the whole community's needs.

Meaningful engagement takes creativity and commitment to reflect on the who, what, when, where, and how of public participation so all stakeholders can contribute. Organizers often complain they have low attendance at public meetings because they planned them at their convenience and not with their stakeholders in mind. To be effective, meet people where they are—practically as well as symbolically. For example, farmers work from sunup to sundown, not from 9 to 5. Their availability is seasonal, affected by weather and what kinds of crops and/or livestock they raise. To engage them, consider leveraging an existing event, or scheduling a twilight meeting in a familiar place during winter months.

You can use any of several reliable approaches to solicit input, share information, and engage community members. *Surveys* are a good way to gather basic information that also can be used to assess conditions and trends. Individual *interviews, listening sessions,* and *focus groups* provide opportunities to relay as well as gather information.

While valuable, these approaches do not strengthen community connections and fall short of fully engaging people in the planning process. Deeper forms of participation bring people together through sustained dialogue. These include various types of *public meetings and forums, charrettes,* and approaches like *World Café*, which can be pursued together or independently. They can include visual techniques and storytelling and require going out into the community instead of inviting the community to come to you.

Public meetings or *forums* foster collaboration and action. They also can educate. Organizing at least two but preferably three or more forums helps guide and inform the planning process. The first could introduce the process and could be used to ground truth data, lift up community values and assets, and generate a shared vision and goals. The second could finalize goals and objectives, revisit outstanding issues, fill in gaps, and bring more people to the table to generate ideas, actions, and recommendations. The third could get feedback on a draft plan, ensure community support, and a path toward implementation.

Public forums are most effective when they are interactive, built around participation, and supported with background materials. Creating a website helps foster a responsive, transparent and collaborative process.[7] Your planning team can decide what kind of meeting format best serves your interests: in person, virtual, or a hybrid combination. For example, meeting in person is best for relationship building but is less accessible. On the other hand, online won't work in places with weak internet access or with populations who do not use the internet. Table 3.2 outlines pros and cons of different approaches.

Table 3.2 In Person or Virtual: Pros and Cons

Format	Pros	Cons
In person	Human connection	Hard to schedule
	Relationship building	Meeting costs
	Spontaneous interactions	Requires transportation
	Body language	Weather dependent
Virtual	Often accessible to more people	Relies on internet access and bandwidth
	Broader reach	Participants must have and know how to use technology
	Easier to record and share after the event	Hard to hold attention and build relationships
	No need for transportation	Hard to work through conflicts
Hybrid	Most flexible	Most complicated and expensive to deliver
	Most accessible	Hard to manage in-person and virtual participants
	Broadest reach	Requires more personnel to manage both audiences
	Reduced transportation requirements	Organizers must have high level of technical savvy

Charettes typically are used to engage stakeholders in the physical design of a planning or community project. They tend to be intense, multi-day events that are carefully designed to promote shared priorities and solutions to problems, which might result in conflicts between residents, developers, and government officials. Instead of working toward solutions over weeks or months or years, they immerse people in a problem for an uninterrupted period of time. Like other engagement strategies, they include cross-sectoral participants and engage key stakeholders from beginning to end in a collaborative process.[8]

World Café is a flexible approach to foster collaborative dialogue, active engagement, and generate ideas for action. It involves seating four or five people together and at least three rounds of progressive, short (usually 20-minute) conversations about issues of common interest or concern. Participants can write, draw, doodle, or otherwise express their ideas on various surfaces, such as placemats, tablecloths, large index cards, or 5 x 7 Post-it notes. Because it offers so many forms of self-expression, this can be a good approach for people with limited literacy or where English is not their first language. To foster cross-pollination of ideas, after each round, participants travel to other tables while one person remains to host the next round. At the end, people share discoveries and insights with the whole group.

Whatever approach you choose, tailor it to the community. Find out as much as you can about residents and respect their culture. Dress appropriately. For example, wear sturdy, weather-proof shoes to a farm. Business casual usually is suitable for a community meeting, but when meeting with local officials or legislators, consider formal business attire. Practice empathy and compassion to build trust through words, tone of voice, and body language. Listen to understand and use inviting—rather than directive—language.

Pay attention to local dialect, offer options and choices, and create opportunities for joint problem solving. If some have low literacy rates or language barriers, adjust your approach. Toward these ends, it helps to use plain language to share meaning and build trust.

People use the same words to mean different things. A submarine sandwich is a hoagie in Philadelphia, a hero in New York, a Po' Boy in New Orleans, and a grinder in New England.

In terms of agriculture, some people only consider large and/or commodity farms "real" farms, limiting engagement with the whole farm community. Or in a more granular example, a "cash deal" is a real estate transaction that does not involve financing. Not knowing the term, a first-time farm buyer with limited English language came to their real estate closing with a bag full of bills and coins instead of a cashier's check or electronic funds transfer. It helps to define terms and use familiar language, and especially when engaging with new Americans, to bring translators—not just to translate words but also cultural nuances and differences.

Beyond civic engagement, six key steps lead to plans, policies, and programs that support sustainable and resilient food systems:

1. Assess conditions and trends affecting food and agriculture.
2. Envision a future based on community values that result in shared goals.

3. Generate recommendations to move forward priorities and achieve goals.
4. Develop and vet the plan.
5. Implement actions: policies, programs, public investments, personnel, and so on.
6. Evaluate progress, adjust tactics.

These steps have a basic trajectory but do not need to follow a prescribed order. This is why the graphic is a spiral rather than a line or a circle. For example, collecting data can inform goals, but once goals are established, you may need more data to inform priorities and recommendations (see Figure 3.1).

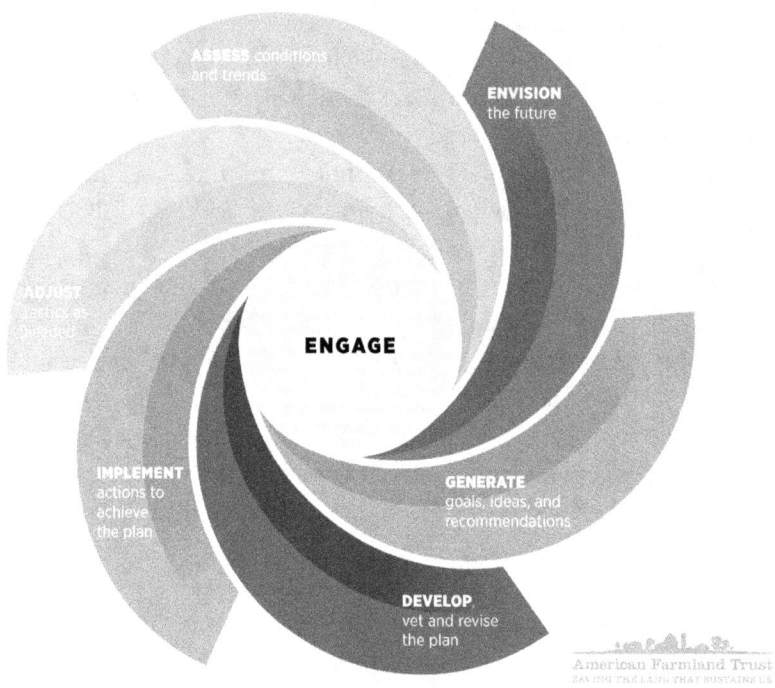

Figure 3.1 Civic engagement is at the heart of an inclusive food systems planning process that assesses trends and conditions, envisions the future, and generates goals and recommendations.

Assess Trends and Conditions

Data illuminates trends and conditions. It helps identify assets, opportunities, and challenges to specific sectors as well as entire food systems. You can collect *primary data* from surveys and interviews and use federal, state, academic, and local sources for *secondary data* to assess food insecurity, production capacity, environmental factors, and so on. Gathering data can be expensive, so it is important to prioritize needs based on budget as well as community priorities. If you have relationships with a college or university, you can tap talented students to help gather and analyze data through an internship, independent study, studio, or planning practicum.

A *SWOT analysis* assesses strengths, weaknesses, opportunities, and threats. Examining both internal and external factors, it is a widely used approach to understanding business or community conditions. It can help you identify issues and inform community discussion. To take an asset-based approach, it helps to emphasize resources while recognizing limitations and vulnerabilities. Multiple datasets, metrics, and analytical tools—including qualitative analysis can help inform your assessment.

Common ways to gather primary data include surveys, stakeholder interviews, focus groups, and listening sessions. Survey information can be collected in a variety of ways depending on your resources and the nature of the community: via internet, telephone, and/or snail mail. *Dot Poster Surveys* are a quick, accessible, and inexpensive way to collect information in public settings. Respondents use sticky dots to answer questions on a large poster board instead of filling out questionnaires or being interviewed.[9]

Lots of secondary data is available, so take the time up front to choose sources that are reliable, relevant, widely available, and can be used to identify trends over time. Local sources include community plans, tax maps and records, aerial photography, and GIS data available through government entities, universities, and/or planning authorities. State sources come from departments of agriculture, economic development, health and human services, and so on. Land-grant universities generally have relevant reports and information. In some states, they support MarketMaker, a searchable database that serves as a virtual infrastructure to connect food and agriculture businesses through the supply chain.

Federal data is widely available and reliable. Some sources are collected or updated annually, others only every five or ten years. However, even the most common and trusted sources have limitations so should be ground-truthed locally. Further, each source has its own definitions and collection methods, so if you combine data from different sources, it is important to be clear about the differences. See Table 3.3 for common go-to sources of federal data for food systems assessments.

Table 3.3 National Sources of Data to Inform Food Systems Assessments

Source	Type of Data	Website
American Community (ACS) Survey	Provides annual U.S. Census Bureau data about people, jobs, education, homeownership, and more. Includes demographic, social, economic, and housing profiles, plus population profiles for many race, ethnic, and ancestry groups. It also includes a series of tables across 86 key variables, including tables comparing counties or congressional districts.	www.census.gov/programs-surveys/acs
Bureau of Economic Analysis (BEA)	Publishes comprehensive U.S. Department of Commerce data from 1969 forward at multiple jurisdictional levels, including economic regions, metro and non-metro areas, cities, states, counties, and national. Datasets include personal income, population trends, and food systems concerns, such as farm income and expenditures, and SNAP receipts.	www.bea.gov/
Bureau of Labor Statistics	Provides annual consumer spending and consumption data on many things, including food by demographic categories as well as geographic region from 1989 to the present.	www.bls.gov/cex
USDA-NASS Census of Agriculture	The most extensive source of data on U.S. agriculture over time, it includes information by county, state, and U.S. territories and has query tools and reports to dig into that data by Congressional districts and zip codes. Data is available on number of farms by size and type, inventory and values for crops and livestock, producer characteristics, land in farms, conservation practices, commodity and marketing data, local food marketing, and more.	www.nass.usda.gov/Publications/AgCensus/2017/

(Continued)

Table 3.3 (Continued)

Source	Type of Data	Website
North American Industry Classification System (NAICS)	Provided by the Census Bureau, NAICS is the standard used by federal and other statistical agencies to classify businesses for the purpose of collecting, analyzing, and publishing statistical data related to the U.S. business economy.	www.census.gov/naics/
USDA Economic Research Service	ERS publishes many reports and datasets relevant to food and agricultural sectors, including on farm production and the farm economy, food and nutrition, food choices, and the rural economy.	www.ers.usda.gov
Soil Survey Geographic Database (SSURGO)	The SSURGO database contains a century of soils information that can be displayed in tables or as maps and is available for most areas in the U.S. and its territories. Datasets consist of map data, tabular data, and information about how the maps and tables were created. Maps can be viewed in the Web Soil Survey (WSS) or downloaded in ESRI® Shapefile format.	https://data.nal.usda.gov/dataset/soil-survey-geographic-database-ssurgo

Various assessment tools support food systems planning. *Asset mapping* is a good tool to identify resources. It can map physical infrastructure like the locations of farms, food retail, processing infrastructure, and waste disposal. Or it can map relationships across the food and agriculture community. *Community food assessments* are a systematic approach to analyzing food system needs and assets. *Cost of Community Services studies and fiscal impact analyses* evaluate a jurisdiction's revenue generation in relation to its costs of providing public services and facilities. *Foodshed analyses* focus on food consumption by exploring past, present, and/or future sources of food for a specific place. *Carrying capacity* and *foodprint* studies investigate the ability of a specific place to feed people, often based on different diet scenarios. Finally, *Economic Impact Assessments* evaluate multipliers or spin-offs to explore total economic impacts—or economic ripple effects of a specific entity or industry—like agriculture (see box).

> ### HOW TO DETERMINE AN ECONOMIC MULTIPLIER
>
> Multipliers refer to economic factors that, when applied, amplify economic impacts. They often are used to provide estimates of employment and output effects of farm and food production to capture indirect and induced impacts as well as direct impacts of economic activities in the sector.
>
> - IMPLAN is a widely used input-output modeling software that can be customized to assess food systems initiatives or the total economic contribution of agriculture and/or a food system to a local economy.
> - USDA's *Local Food Economic Development Toolkit* provides extensive information on secondary data sources as well as multipliers and includes a chapter on the use of IMPLAN.
> - ERS has a website with instructions on how to develop Agricultural Trade Multipliers that provide annual estimates of employment and output effects of trade in farm and food products on the U.S. economy.
> - Some land-grant universities provide information on how to develop multipliers. Two examples are:
>   - New Mexico State University has a webpage on Income Multipliers in Economic Impact Analysis.
>   - The University of Tennessee Extension Institute of Agriculture published "A Primer in Economic Multipliers and Impact Analysis Using Input-Output Models."

Envision a Desired Future and Set Goals to Achieve It

Another early and essential step is to bring community members together to create an aspirational vision of the future. Visioning often is combined with identifying shared values and goal setting. Other times, goal setting occurs later in the process. Whatever the order, these activities are best facilitated in person to build trust between stakeholders. Even people from very different cultural and political persuasions share values and goals. Expressing them together establishes common ground.

Schedule meetings for a time and place convenient to residents. Reach out to representatives from different sectors and populations and ask them to identify people to ensure key stakeholders are included. If possible, serve appropriate refreshments and—especially for an evening or weekend meeting—offer childcare.

Visioning is a creative, judgment-free process. It often starts by brainstorming a list of ideas. After that, the group can prioritize, evaluate, and narrow down ideas to reach agreement. Small groups can work together to brainstorm a vision together, or individuals can write down their visions and then share with the group. Sharing can occur in many ways, such as posting sticky notes on a wall, reading out loud, working in small groups, or compiling into a written document that is brought back to the group. This process also works to establish shared values.

Appreciative inquiry is an asset-based approach. It uses questions and dialogue to help participants generate positive ideas.[10] It also can be used to uncover community assets, strengths, and opportunities. *Scenario planning* is another approach to visioning. Here participants develop alternative visions based

Example of Goals from a Food Systems Engagement Meeting

GOALS		GOALS CONT.	
1. INCREASE INFRASTRUCTURE INVESTMENT (INCL TRANSPORT)	3 ✓	11. ADAPT TO THE CHANGING FOOD SYSTEM THROUGHOUT THE PROCESS OF ACHIEVING GOALS TO KEEP EFFORTS DYNAMIC	1 ✓
2. PASS THE ZONING ORDINANCE FOR URBAN AG	4 ✓		
3. ENGAGE ALL FOOD SYSTEM STAKEHOLDERS (INCL CON AGRA, LOCAL GOV'T, FNDNS, ETC.)	5 ✓	12. INCREASE THE AMOUNT OF LOCAL FOOD PURCHASED BY INSTITUTIONAL BUYERS	7 ✓
4. MAKE FOOD SYSTEM MORE UNDERSTANDABLE TO GENERAL POPULATION + DECISION-MAKERS	9	13. INCREASE ACCESSIBILITY TO HEALTHY, LOCAL FOOD THROUGHOUT THE COUNTRY	2 ✓
5. INCREASE EDUCATIONAL INITIATIVES ABOUT THE FOOD SYSTEM + ALL THAT ENTAILS	5 ✓	14. INCREASE LOCALLY FINANCED PRODUCTION + PROCESSING TO LOCAL MARKETS	1 ✓
6. STREAMLINE FARM SUCCESSION	1	15. ESTABLISH A MECHANISM FOR	1

Figure 3.2 Example of goals from a food systems engagement process.

upon different conditions. This allows them to imagine what the future will look like under "business as usual" and more creative scenarios. Whatever approach you choose, it helps to start with a worksheet with a list of key questions to focus attention.

Where visioning is broad and aspirational, goal setting is grounded in specifics. Writing down goals during an interactive process helps facilitate community agreement to ensure they are shared. To avoid having too many goals, it helps to identify overarching goals with concrete objectives. Communities often strive to set "SMART" goals, which are **S**pecific, **M**easurable, **A**chievable, **R**ealistic, and **T**ime-bound.

The core planning team can pull information together in a draft of a shared statement, which must be vetted before it is finalized. Ensure everyone's voice is heard and build on this to identify common themes and to set and prioritize goals and objectives that can be achieved in a realistic time frame.

Generate Recommendations to Advance Priorities and Achieve Goals

Once the community has agreed on a vision and drafted goals, it is ready to propose strategies and actions to achieve them. These should address assets and opportunities and be based on trends and conditions. This is a good time to engage the residents to ensure the actions proposed meet their needs and to build a base of support for implementation.

The best plans tie recommendations to goals and objectives, so each solution proposed is a step toward achieving a stated priority. Important considerations include urgency, cost, and resource availability. Strong recommendations include low-hanging fruit—easy wins and short-term strategies, as well as more complex mid- and long-term strategies, like policy changes.

It helps to have a timeline, assigned roles and responsibilities, and accountability measures. As part of this, it is wise to identify funding to move recommendations forward. Funding may come from private sources like foundation grants, as well as federal, state, and local sources.

Toward these ends, it helps to spend time addressing the five Ws:

- **Who will do What** (how and for whom);
- **Why** is it important; and
- **When and Where** will they do it?

Also, consider dependencies—how one action depends on another. For example, implementing a farm-to-school program depends on certifying farmers in Good Agricultural Practices (GAP). Farmers must be trained and GAP certified before the school year for the school to buy their products.

Finally, develop targets, benchmarks, and indicators to measure progress. *Targets* are aspirational and should be aligned with the vision. *Benchmarks* reflect existing conditions and are used to measure progress toward specific changes, objectives, or goals. *Indicators* are measures used to evaluate progress.

Develop and Vet the Plan

Once these steps are complete, it is time to pull everything together and draft a plan that reflects the outcomes of the processes you choose. This usually is led by the project manager with input and support from the core team and advisors. Basic elements include background information based on the assessment of current trends and conditions, likely including maps, benchmarks, and a SWOT analysis. They include the vision, goals, and objectives the community set as well as the recommended strategies and actions to achieve them. Since plans should be transparent, describe the people, partners, and processes involved in creating the plan and including appendices that reflect public input. And they should explain the process for evaluating the plan, reporting on its implementation, and adjusting over time.

Once drafted, it is important to bring the draft plan back to the community for public comment and input. Make sure to incorporate constructive feedback and keep people informed of the timeline for completion, how it will be made available to them, and their roles.

Implement Strategies and Actions

A plan is only as good as its implementation. So once drafted, work through who will take what actions, when, and how. Actions may include new personnel, partnerships, programs, and both public and private investment. They also include creating new—or eliminating outdated—policies.

It helps to look at what other communities have done to address goals and challenges that are similar to yours. However, there is no silver bullet; one size does not fit all. The second half of the book introduces a toolbox of program and policy options you can adapt to meet your community's needs.

Implementation happens over time. Metrics inform how to modify strategies to get where you want to go. Measuring results show if there is a need to adjust tactics and targets to achieve goals. Metrics can be based on the plan's initial benchmarks, and indicators and can be changed if new information is useful or becomes available.

> **METRICS TO MEASURE PROGRESS**
>
> Led by staff from the state economic development authority, Vermont Farm to Plate is a holistic food systems plan with 15 strategic goals organized into four categories:
>
> - Sustainable Economic Development;
> - Environmental Sustainability;
> - Healthy Local Food for All Vermonters; and
> - Racial Equity.
>
> Each goal has clear indicators of progress, which are tracked and evaluated over time. Some contain known targets, others need to establish baselines. Some include complex analyses. For example, Goal 1 is to increase food system economic output by $3 billion by 2030, measured using Economic Census and NASS production data in five-year intervals. The Farm to Plate website has a Data and Outcomes page where users can follow progress.
>
> Various other metrics are used to inform planning and community action. These three examples come from organizations that offer workshops and training to use their tools:
>
> - The Center for Whole Communities created a Whole Measures to assess relationships between land, people, and communities. A flexible approach, it is accessible and adaptable for different uses and users.
> - The Fiscal Policy Studies Institute developed a Results-Based Accountability™ framework to help communities measurably improve the well-being of their citizens.
> - Yellow Wood Associates has developed a "You Get What You Measure®" approach to strategic planning. Described as a values-based approach, it incorporates metrics to use as a framework for evaluation.

PATHWAYS TO PLANNING FOR FARMS AND FOOD IN WAYNE COUNTY, OHIO

Figure 3.3 Wayne County is a rural county with a strong agricultural base located roughly halfway between Cleveland and Columbus, Ohio.
Source: Wayne County Planning Department.

With a dynamic and diversified farm economy, Wayne County leads Ohio in dairy production and local food sales and supports two of the state's 12 produce auctions.[11] Its Ag Success Team (AST) is an innovative collaboration of farm interests created in 2004 by a now retired county commissioner as a loose network of local government, Extension, Farm Bureau, Farm Credit, Soil and Water district, and other county institutions. The AST meets monthly to network and identify issues affecting county agriculture. Farming is further supported by a vibrant processing sector with several local cheese makers, dairy cooperatives, and meat processors, plus national firms like Daisy Brands and Smucker's. Still, the county's food system is threatened by external factors, including development pressure, escalating land values, farm consolidation, and a disconnect between local production and consumption.

In 2019 the county adopted an update to its comprehensive plan, *Wayne County Onward*. Its vision and goals include abundant farmland;

preserving heritage, culture, and natural resources; and supporting a diversified local economy. Typical of many rural plans, it seeks to preserve farmland without connecting farms to economic development or food. And it largely ignores the county's Amish and other Plain communities, which play a significant role in the county's agriculture.

As part of a USDA-funded project, Wayne County–based Ohio State University (OSU) faculty and Extension educators led a community engagement process to strengthen these connections. The process centered on building rural wealth through food and agriculture and preparing the county for future challenges. Extension leaders convened a core planning group that included American Farmland Trust, Amish farmers, county Farm Bureau representatives, local food organizations, OSU County Extension, and the OSU-Ohio Agricultural Research and Development Center (OARDC).

The planners launched the process with a community forum: "Economic Opportunities Through Food and Agriculture." They used Ketso—a reusable, hands-on community-engagement toolkit[12]—to develop an invitation list. The list included diverse types of farmers, both Amish and "Englisch" (as the Amish refer to residents who are not from a Plain sect), and individuals from the business, finance, government, health, nonprofit, and educational sectors. Then the team worked together to create the program, conduct outreach, and plan a menu comprising mostly local food.

The event was free, but registration was required. Most participants signed up online. To accommodate the large Plain community, team members helped register people who did not use the internet. They also used the registration list to assign people to tables to ensure interactions across sectors and between people who did not usually talk to each other. To emphasize the value of the regional food economy, 90 percent of the food that was served at the March event was sourced from Wayne and its neighbors, Holmes and Ashland counties.

The forum was the county's first cross-over event that included diverse food and agriculture perspectives. It convened 75 key stakeholders from food, agriculture, economic development, and local government to identify opportunities to strengthen the county's food and agriculture economy. The program opened with a review of local trends, conditions, and assets. This was followed by presentations from the U.S. Federal Reserve and USDA Rural Development to highlight economic development opportunities through food and agriculture.

Most of the day focused on interactive small group activities. The first was a *mapping activity* where participants worked at pre-assigned tables to identify the county's food system assets. Each table had six to eight people from various sectors and points of view. They were given a county map and a set of stickers, which represented assets from farms to processing facilities to supermarkets to funders to people. Participants were asked to introduce themselves and their interest in the county's food system, discuss local farm and food assets, and then put stickers on the map to reflect the assets' physical locations. Once finished, someone from each table pinned their map on a cork board for others to see.

The second activity was a *small group brainstorming activity* to identify and prioritize what was needed to grow economic opportunities through food and agriculture. The planners provided ground rules to facilitate the conversations and examples of possible next steps. Each table was asked to discuss three things and record them on a flip chart:

1. What surprised them and/or what they observed from the mapping exercise.
2. Ideas for growing the county's food and farming economy.
3. Which two ideas to pursue as concrete next steps and how to make them "SMART."

Groups had half an hour for discussion. Then each table reported out to the whole group.

Based on very positive feedback from the forum, the planning group expanded to include county economic development staff. Together, they organized a follow-up two-day engagement event six months later to showcase the economic contributions of the county's diverse value-added food and agriculture enterprises. Participants included members of the finance and banking community, county planning, economic development and emergency management staff, the county career center, migrant justice staff, nonprofits, farmers, and a state representative.

The second event started with a narrated bus tour that included stops at a local meat processor, a produce auction, a fruit farm, a diversified farm with a farm store, and a local food coop. Participating businesses received a $200 stipend and an OSU Certificate of Appreciation and Participation, which they subsequently hung proudly in their stores. To build bridges across the Amish and Englisch communities, participants enjoyed an Amish wedding dinner at a local Amish U-Pick fruit farm.

Participants learned about the county's value-added aggregators and processors and the important role they play in the local economy. Each received a gift bag that included a $10 gift certificate to the local food co-op, coupons for participating businesses, and samples of local food products.

The second day included a luncheon presentation and a lively afternoon workshop for community leaders to build on and expand opportunities identified at the prior events. It was held in a historic barn in a town neighboring the city of Wooster. Again, the menu was largely locally sourced. During the event, the P2P team reflected on both events and encouraged participants to do so as well. Amish and Englisch members identified common issues and debunked misconceptions.

The next day, the planners joined an AST meeting and brought along Plain community and direct marketing farmers who generally did not attend. Leaders who had participated in the prior days' events shared observations with those who had not, identified common challenges and opportunities, and built greater cross-cultural understanding and energy for action.

Over the next few months, The AST shifted from a loose network to a working-group model to promote new markets, land use policy, farmer support, and food initiatives. It recruited new members to better represent the totality of the agricultural community and embarked on a process to rewrite its mission and work more closely with county commissioners to achieve shared goals. County commissioners created a new position in the planning department to carry the work forward and hired a core member of the planning team to create a new plan to support rural wealth creation through diversified food and agriculture.

Discussion Questions

1. Who needs to be included in your planning process?
2. How will you reach and engage them?
3. What data will you use to assess trends and conditions?
4. Who could you ask for help?

Notes

1. "Principles of a Healthy, Sustainable Food System," *American Planning Association*, accessed March 6, 2023.
2. Global Alliance for the Future of Food, "Principles for Food Systems Transformation: A Framework for Action," *Global Alliance for the Future of Food*, June 2021.
3. Lionel J. Beaulieu, "Mapping the Assets of Your Community: A Key Component for Building Local Capacity," *Southern Rural Development Center*.
4. Cornelia Butler Flora, Jan L. Flora, and Susan Fey, *Rural Communities: Legacy and Change*, 2nd ed. (Boulder, CO: Westview Press, 2004). The book is the companion text to a PBS college-level telecourse and television series entitled "Rural Communities: Legacy and Change" which portray the experiences of fifteen rural communities across the U.S.
5. Jill K. Clark et al., "Fail to Include, Plan to Exclude: Reflections on Local Governments' Readiness for Building Equitable Community Food Systems," *Built Environment* 43, no. 3 (September 1, 2017): 315–27.
6. U.S. Census Bureau, "State Population Totals and Components of Change: 2020–2022," Census.gov, accessed January 29, 2023.
7. To learn more about conducting public forums with farmers: Maria Pippidis et al., *Engaging Communities Through Issues Forums: A How-To Guide for Onsite and Online Community Engagement* (Kansas City, MO: Extension Foundation, 2022).
8. To learn more about conducting a charette: Bill Lennertz and Aarin Lutzenhiser, *The Charrette Handbook*, 2nd ed. (New York: Routledge, 2017).
9. Larry Lev and Garry Stephenson, "Dot Posters: A Practical Alternative to Written Questionnaires and Oral Interviews," *Journal of Extension*, Tools of the Trade 37, no. 5 (October 1999).
10. Muriel A. Finegold, Bea Mah Holland, and Tony Lingham, "Appreciative Inquiry and Public Dialogue: An Approach to Community Change," *Public Organization Review* 2, no. 3 (September 1, 2002): 235–52.
11. USDA National Agricultural Statistics Service, "2017 Census of Agriculture, County Profile: Wayne County, Ohio," 2017.
12. For information on Ketso, "Workshops & Engagement | Ketso | Ketso Workshops," Ketso, accessed July 2, 2023.

Section II

4

FEDERAL POLICIES THAT AFFECT FOOD SYSTEMS

Chapter Summary

A wide range of federal agencies and policies affect food systems, from the Clean Water Act to the Farm Bill to the Federal-Aid Highway Act. This chapter provides an overview of major agencies and policies that influence food systems and—because of their significant roles—goes into greater depth about USDA and the Farm Bill. It also sets up subsequent chapters that offer promising cases and examples of what state and local governments can do to address food systems challenges and implement food systems plans.

***

Plans must be implemented, at least in part, with programs, policies, and public investments. The federal government largely plays an indirect role in land use planning, as it owns and manages about 30 percent of the nation's total surface area—and nearly half of western states.

It plays a more direct role in food, agriculture, and environmental policy. Driven by competing interests, its agency-by-agency approach has led to uncoordinated, and in some cases contradictory, actions—even within agencies. It is beyond the scope of this book to unravel this. But some groups and agencies have made the case for a national food strategy.

The U.S. General Accountability Office has called for a government-wide approach to address fragmentation in the federal food safety oversight system.[1] The Harvard Food Law and Policy Clinic partnered with Vermont Law School's Center for Agriculture and Food Systems to create a food strategy blueprint to inspire more coordinated federal food and agricultural law and policymaking.[2] Given the lack of coordinated federal policy, it is important to understand the underlying impact of federal policies on food systems efforts.

The Federal Role in Land Use Policy

The federal government owns and manages about 650 million acres of land in the U.S. Nearly a third of this is controlled by two agencies, the Bureau of Land Management (BLM) and USDA's Forest Service. They both manage land for multiple, sustainable uses, which include grazing.[3] But all federal land management agencies must balance competing interests from state, local, and tribal governments both to increase the use and protection of these resources.

Outside of the lands it owns, the federal government lacks power to affect land use planning, but policies often intended for other purposes influence who uses land and how and where those uses occur. For example, the interstate highway system and federal tax regulations, like the mortgage interest deduction, spur sprawl.[4] Further, the U.S. has signed treaties that influence land use on tribal lands, and it provides financial incentives to state and local governments for development projects.

One of its largest influences is the National Interstate and Defense Highways Act of 1956 (Federal-Aid Highway Act), which created the interstate highway system. With the subsequent Highway Act of 1958, it drove suburban development. Fueling farmland conversion, it fundamentally changed land use patterns across the U.S. As superhighways cut through urban neighborhoods, they demolished affordable housing and spurred the growth of suburbs. By the turn of this century, sprawl had become a dominant feature of the American landscape.[5]

The impacts of highway development on farmland continue to be felt today. Between 2017 and 2021, the Department of Transportation (DoT) alone generated a third of federal project proposals to convert agricultural land[6] (see Figure 4.1).

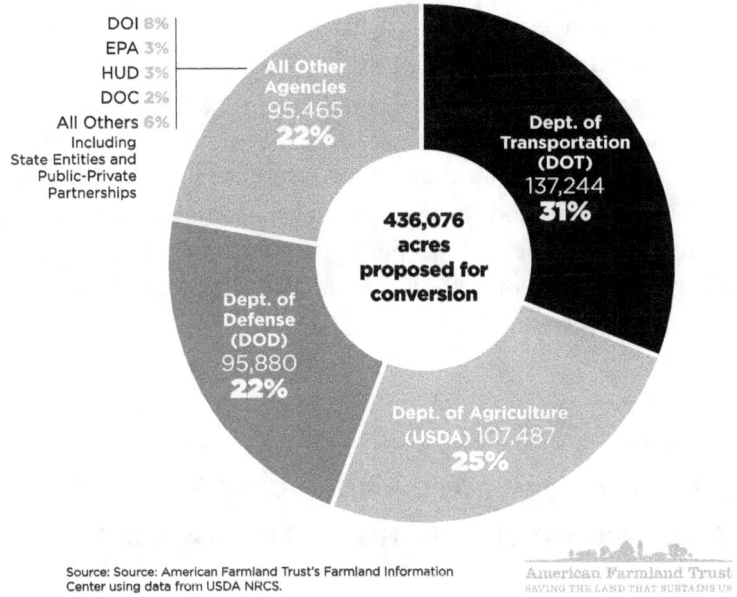

Figure 4.1 Congress enacted the Farmland Protection Policy Act (FPPA) in 1981 to minimize the extent to which federal programs contribute to the unnecessary conversion of farmland to non-agricultural uses. NRCS tracks farmland conversion proposed by federal agencies. After DoT, it found USDA is the second greatest driver of conversion, followed by the Department of Defense.
Source: American Farmland Trust Farmland Information Center.

USDA and the Farm Bill

The Farm Bill is complex legislative package that is periodically renewed (generally every five years). It authorizes policies and programs and funds many of them through direct or mandatory funding, which is authorized through the bill, or through discretionary funding, which must be approved through annual Congressional appropriations. Programs like crop insurance have permanent authority. Others require reauthorization, most notably nutrition assistance and commodity support programs.

Historically, the Farm Bill focused on commodity program support for a handful of staple crops: corn, soybeans, wheat, rice, peanuts, dairy, sugar, and cotton. Over time it expanded to support nutrition, conservation, crop insurance, rural development, and specialty crops, among other things.[7] Specialty crops include fruits and vegetables, tree nuts, horticulture, floriculture, and nursery crops.

The *Agricultural Improvement Act of 2018* is a sweeping bill with 12 Titles (see box). Due for reauthorization in 2023, the bill was extended through September 30, 2024. The nutrition Title receives the most support, followed by crop insurance, commodity programs, and conservation. Less than 1 percent of direct funding goes to the other eight titles. Thus a bit of a misnomer, the Farm Bill is really more of a food bill. It is also a significant piece of environmental legislation.

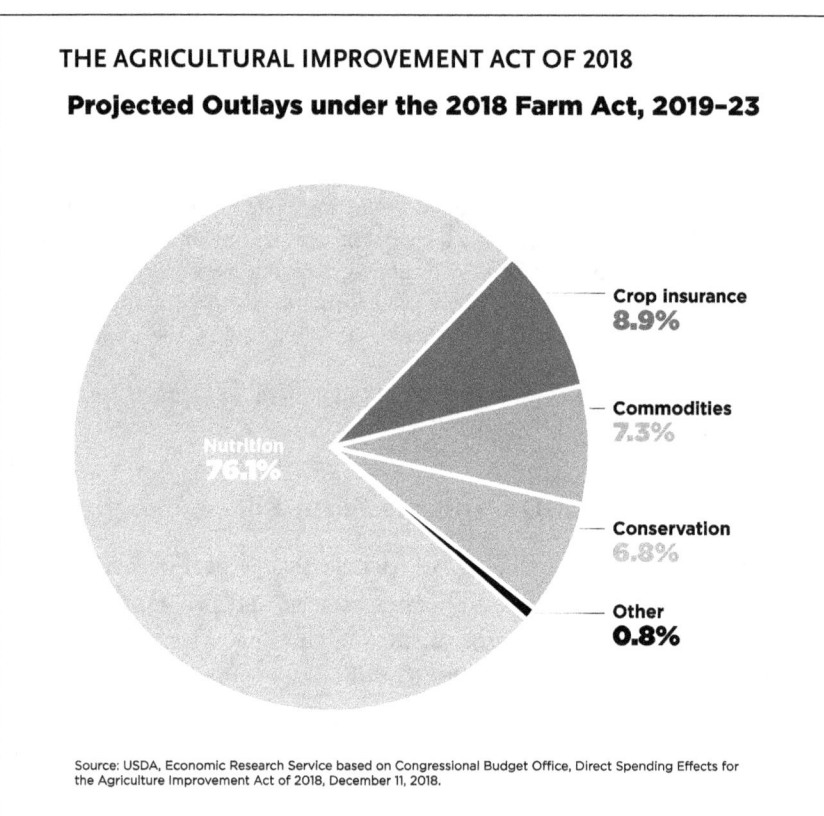

Figure 4.2 Most Farm Bill spending is for nutrition assistance.

1. **Title I, Commodities**: Supports major commodity crops and disaster assistance. Major commodities include corn, soybeans, wheat, rice, peanuts, dairy, and sugar.
2. **Title II, Conservation**: Supports environmental stewardship, improved agricultural land management, and conservation easement programs. These include both working lands and land retirement programs.
3. **Title III, Trade**: Supports agricultural exports and international food assistance.
4. **Title IV, Nutrition**: Provides nutrition assistance to children and low-income households and includes 15 programs. SNAP is by far the largest.
5. **Title V, Credit**: Provides government loans and guarantees to producers to buy land and operate farms and ranches.
6. **Title VI, Rural Development**: Supports rural housing, community facilities, business, and utility programs through grants, loans, and loan guarantees.
7. **Title VII, Research and Extension**: Supports agricultural research and extension programs to expand academic knowledge and to support food safety and nutrition, farmers and ranchers, farming practices, and production systems.
8. **Title VIII, Forestry**: Supports USDA Forest Service's management programs.
9. **Title IX, Energy**: Promotes farm and community renewable energy systems through various programs, including grants and loan guarantees.
10. **Title X, Horticulture**: Supports specialty crop production, USDA-certified organic and locally produced foods, and authorizes a regulatory framework for industrial hemp.
11. **Title XI, Crop Insurance**: Supports risk management through a Federal Crop Insurance Program.
12. **Title XII, Miscellaneous**: Includes programs and assistance for livestock and poultry production, support for beginning farmers and ranchers, and other miscellaneous and general provisions.

Food and Nutrition

President Johnson signed the Food Stamp Act of 1964 as a cornerstone of his War on Poverty. Over the next decade, states and counties added it to their portfolio of services for low-income residents. By 1974 it was available nationwide. Currently called the Supplemental Nutrition Assistance

Figure 4.3 SNAP logo.

Program (SNAP), it is the centerpiece of the nutrition Title. Reflecting pandemic demand and administrative adjustments, the Congressional Research Service projected that for 2022, nutrition comprised 84 percent of the Farm Bill baseline.[8]

SNAP is administered by states, which set eligibility criteria. Recipients apply where they live and must meet income and resource requirements, which are updated annually. Benefits are delivered monthly on an Electronic Benefit Transfer (EBT) card, which works like a debit card and can be used to buy groceries at authorized food stores and retailers. Other significant Farm Bill food programs include:

- The *Emergency Food Assistance Program*, which provides food commodities through states to food banks and pantries;
- The *Commodity Supplemental Food Program*, which provides supplemental food packages primarily to low-income elders; and
- The *Senior Farmers' Market Nutrition Program*, which provides vouchers to low-income seniors to buy fresh produce at farmers' markets and other direct-to-consumer venues.

In lieu of SNAP, the *Food Distribution Program on Indian Reservations* provides food to low-income households on reservations and to Indigenous families living in or near designated areas in Oklahoma.

Two grant programs support food access and security. *Community Food Projects* (CFPs) fund competitive grants to food service providers from public agencies, tribal organizations, and nongovernmental organizations (NGOs) to address food and nutrition insecurity, especially among vulnerable populations. First authorized in the 1996 Farm Bill to meet the needs of low-income individuals, CFPs support community outreach, food distribution, and local responses to meet state, tribal, and local food access and agricultural needs.

The *Gus Schumacher Nutrition Incentive Program* (GusNIP) is a competitive grant program authorized through 2023. It funds projects to help low-income consumers buy fresh produce by providing incentives at SNAP points of purchase or providing produce prescriptions to SNAP/Medicaid participants. It includes three programs. The *Nutrition Incentive Program* (NIP) provides point-of-purchase incentives for fruits and vegetables to SNAP beneficiaries and other income-eligible consumers. The *Produce Prescription Program* (*Veggie Rx*) demonstrates and evaluates the impact of fresh fruit and vegetable prescriptions on human health. The third program offers grants for training, technical assistance, evaluation, and information center cooperative agreements. Together, GusNIP brings stakeholders together from across the food and health care sectors to

Figure 4.4 The Child Nutrition Act authorizes all federal school meal and child nutrition programs, including the Special Supplemental Nutrition Program for Women, Infants, and Children (WIC).
Source: USDA. Photo by Preston Keres.

foster understanding and work together to improve public health, facilitate growth in underrepresented communities and geographies, and aggregate data to identify and improve best practices on a broad scale.

The *Child Nutrition Act* was signed into law in 1966. It authorizes all federal school meal and child nutrition programs, including the *Special Supplemental Nutrition Program for Women, Infants, and Children (WIC)*. Primary programs include *School Breakfast and Lunch*. All children may participate in a school that operates the programs, but most schools target children certified to receive free or reduced-price meals. Other programs include summer feeding and after-school meals. The 2008 Farm Bill directed the Secretary of Agriculture to encourage institutions operating child nutrition programs to purchase unprocessed, locally grown and raised farm products. This opened the door to a flurry of state and local activity. USDA's 2019 Farm to School Census found that about two-thirds of all school food authorities participated in one or more farm-to-institution activities and that local foods accounted for about 20 percent of overall food purchases, excluding entitlement spending.[9]

WIC provides nutritious foods, nutrition education, and access to health care to pregnant women, new mothers, infants, and children up to age five living in low-income households. Supplementing WIC, the *Farmers Market Nutrition Program* issues coupons to WIC recipients to buy fresh, locally grown fruits and vegetables at participating farmers' markets. FNS awards grant funding to state agencies and tribal organizations to administer the program. Eligible WIC participants receive FMNP coupons on top of regular benefits to buy fresh produce from farmers, farmers' markets, and roadside stands. WIC's *Fresh Fruit and Vegetable Program* promotes nutrition education and offers free produce to children at eligible elementary schools during the school day.

Food Safety

Food safety is governed by a complex system of at least 30 laws administered by 15 federal agencies. USDA Food Safety and Inspection Service (FSIS), the Department of Health and Human Services (HHS), and the Food and Drug Administration (FDA) have primary oversight. While they work together, FSIS and FDA have overlapped authority. For example, the FDA regulates the safety of eggs in the shell, while USDA regulates safety for eggs removed from the shell and runs an egg marketing program for eggs in the shell.[10] FSIS has jurisdiction over meat and poultry, and FDA for everything else.

The 2011 *Food Safety Modernization Act (FSMA)* gave FDA responsibility for enforcing most of the nation's food supply. It was an effort to address the fragmentation of federal food safety oversight, which shifted FDA's focus from responding to foodborne illness to preventing it. FSMA rules affect human food and animal feed, produce safety, third-party certification, foreign compliance, and requirements for food facilities and transportation. FDA's Food Traceability Final Rule establishes additional traceability recordkeeping requirements for anyone who manufactures, processes, packs, or holds foods included on the Food Traceability List (FTL) to allow for faster identification and rapid removal of potentially contaminated food from the market, resulting in fewer foodborne illnesses and/or deaths. As important as FSMA is to preventing foodborne illness, its rules do not consider the ecological impacts of the farm management adjustments it requires. For example, they discourage natural vegetation and compost, which attract pollinators, reduce soil erosion and runoff, improve downstream water quality, and sequester carbon. If FDA-approved fertilizers enter local water ways, they are predicted to have significant environmental and public health consequences that may not be apparent for years.[11]

Agriculture and the Environment

Figure 4.5 A great "roller" moves across the land during the Dust Bowl of the 1930s. Source: USDA.

Congress passed the Soil Conservation Act of 1935 in response to the Dust Bowl—the legendary period when severe dust storms ravaged the Great Plains. Starting in 1932, a severe drought, coupled with soil erosion, caused widespread crop failures. In March 1935, a series of intense and frequent storms swept the Plains and blew dark clouds of fine particle dust all the way to Washington, D.C. President Roosevelt signed the Act in April, creating the Soil Conservation Service (SCS).

The Act led to the creation of local conservation districts. Today nearly 3,000 districts coordinate assistance from local, state, federal, and private sources to develop locally driven solutions to a wide range of natural resource concerns. Established under state law, they are called various things from conservation to soil and water to resource conservation districts or committees.

Over time, SCS's role expanded to include surveys and flood control plans, land use, farm forestry, and water concerns to become USDA's leading private lands conservation agency. In 1994, it was renamed the Natural Resources Conservation Service (NRCS). Today, NRCS and the Farm Services Agency (FSA) administer a complex portfolio of Farm Bill programs incorporated into the conservation title. The conservation title was added to the Farm Bill in 1985, advocated by a newly formed coalition of agriculture and environmental groups, including American Farmland Trust. Initially focused on preventing erosion, Title II has been expanded over the years to include programs that build soil health, enhance farm viability, and protect farmland and ranchland from development. Current programs can be categorized as follows:

Easement programs pay landowners to use conservation easements to limit non-farm development and protect land and wetland resources. NRCS administers the *Agricultural Conservation Easement Program (ACEP)*. ACEP has two components: *agricultural land easements (ALE)*, and *wetland reserve easements (WRE)*. ALE provides matching funds to state and local governments, tribes, land trusts, and other conservation entities to purchase agricultural conservation easements on working farms and ranches. WRE helps private and tribal landowners protect, restore, and enhance wetlands that have been damaged by agricultural uses. Additionally, through its *Wetland Reserve Enhancement Partnership*, NRCS enters into agreements with eligible partners to carry out high-priority wetland protection, restoration, and enhancement, and improve wildlife habitat.

Land retirement programs authorize USDA to pay private landowners for temporary changes in land use and management to achieve environmental benefits. The *Conservation Reserve Program* (CRP) is its signature program. Managed by FSA, CRP provides annual rental payments to remove environmentally sensitive land from farm production and plant species to improve soil and water quality and provide wildlife habitat. Voluntary lease agreements typically last for 10–15 years. CRP has several subprograms.

- *Grassland* CRP protects rangeland, pastureland, and other lands while maintaining them for grazing;
- The *Farmable Wetlands Program* is designed to restore up to 1 million acres of previously farmed wetlands and wetland buffers to improve vegetation and water flow;
- The *State Acres for Wildlife Enhancement* (SAFE) establishes wetlands, grasses, and trees to create critical habitat and food sources to enhance important wildlife populations; and
- The CRP *Transition Incentives Program* (TIP) provides annual payments for two years beyond a CRP contract expiration to sell or lease land to a beginning, veteran, or socially disadvantaged farmer or rancher.

In addition, the *Conservation Reserve Enhancement Program* (CREP) authorizes USDA to enter into agreements with states to target areas to address specific environmental objectives. It typically provides additional financial incentives beyond annual rental payments to eligible partners.

Conservation Cost Share Programs keep private land in production while providing participants with planning, technical assistance, and financial support to defray some of the costs of installing or maintaining conservation practices. NRCS administers two main programs: the *Environmental Quality Incentives Program* (EQIP) and the *Conservation Stewardship Program* (CSP). Combined, they account for more than half of all conservation program funding.

EQIP is NRCS's flagship program to help farmers, ranchers, and forest landowners integrate conservation into working lands. It provides financial and technical assistance to plan and install land management practices that lead to cleaner water and air, healthier soil, and better wildlife habitat while improving agricultural operations. It also supports initiatives that target financial assistance to specific kinds of projects in some states, including for high tunnels to extend the growing season and organic and on-farm energy initiatives.

Figure 4.6 Rice harvest at 3S Ranch, near El Campo, Texas. The ranch used EQIP cost share funding to make efficient use of water on irrigated land.
Source: USDA Media by Lance Cheung

CSP provides financial and technical assistance to producers to maintain and improve existing conservation systems and to adopt comprehensive conservation activities on entire operations. These include crop rotations, cover crops, advanced grazing management, and support to help producers develop a comprehensive conservation plan.

Partnership and grant programs leverage nonfederal funding. Among them, NRCS manages the *Regional Conservation Partnership Program (RCPP)* and *Conservation Innovation Grants (CIG)*. RCPP leverages federal conservation funding for partner-defined projects. Piloted in the 2014 farm bill, it was reauthorized in 2018 as a stand-alone program to fund activities related to USDA programs, including EQIP, ACEP, and CRP. CIG awards competitive grants to state and local agencies, NGOs, tribes, and individuals to implement innovative conservation practices.

Beyond USDA, other federal agencies and several pieces of historic legislation affect agriculture and food systems. The *U.S. Fish and Wildlife Service* and the *National Oceanic and Atmospheric Administration's National Marine Fisheries Service* jointly administer the *Endangered Species Act* of 1973, which protects threatened or endangered fish, wildlife, and plants. All federal agencies are required to comply and ensure their actions will not jeopardize listed species or damage critical habitat. ESA also prohibits agricultural practices from contributing to habitat and biodiversity loss on private lands. Beyond ESA, the Bureau of Land Management (BLM), within the Department of the Interior (DoI), plays an important role in grazing leases. And the Environmental Protection Agency (EPA) plays a multifaceted role protecting food as well as water and environmental quality.

Bureau of Land Management

The *Federal Land Policy and Management Act* of 1976 (FLPMA) authorized the federal government to retain ownership and prevent degradation of most public lands. BLM has a planning process to inventory lands and resources and develop management plans to protect habitat, promote conservation and recreation, and allow energy and mineral extraction. Finally, FLPMA requires federal agencies to coordinate planning and management processes with local and state governments and plans.

Through FLPMA, BLM administers grazing permits and leases, mostly for cattle and sheep, on more than 21,000 allotments. These generally cover a ten-year period and are renewable if the agency determines that the terms and conditions of the expiring permit or lease are being met.[12] The *Public Rangelands Improvement Act* of 1978 requires the Secretaries of both the Interior and Agriculture to develop, update, and maintain an inventory of range conditions and to track the condition of public rangelands. It also established a formula for setting livestock grazing fees, including on rangeland BLM leases to ranchers.

Environmental Protection Agency

The EPA plays several roles in food systems. On the food side, it works to ensure that pesticides used on food meet strict safety standards under the *Food Quality Protection Act* (FQPA). It promotes a systematic approach to the sustainable management of food to reduce food waste and its associated impacts, starting with the use of natural resources and progressing through the life cycle of food to recovery or final disposal, including composting.

Toward these ends, it teams up with USDA through the U.S. Food Loss and Waste 2030 Champions group—businesses and organizations that have made a public commitment to reduce food loss and waste in their own operations by 50 percent by the year 2030.

EPA administers numerous regulations affecting agricultural impacts on air and water quality, pesticide use, storage, and disposal, animal feeding operations (AFOs), land contamination, and so on. The most significant is the *Clean Water Act (CWA)*. Passed in 1972, it regulates quality standards for surface waters and pollution discharges into U.S. waters. CWA gave EPA authority to implement measures such as setting wastewater standards for industries and requiring permits for pollutants discharged from a point source into navigable waters. Many agricultural activities are exempt if they follow best management practices, but as part of the *CWA/Safe Drinking Water Act*, farms must obtain permits for some activities, including discharges from animal feeding operations, pesticide applications that contaminate water, and a series of regulations related to storage, discharge, and use of oil and oil products.

Through *Nonpoint Source Management Program Section 319*, EPA provides grants to states, territories, and tribes for financial and technical assistance, monitoring of nonpoint source implementation projects, and other things. It addresses nonpoint source pollution from excess fertilizers, herbicides, and insecticides from residential as well as agricultural lands, urban run-off, nutrients from livestock, pet wastes and faulty septic systems, and more.

EPA also provides resources to help communities protect the environment, improve health, and strengthen local economies. It encourages renewable energy development on current and formerly contaminated lands, landfills, and integrates smart growth, environmental justice, and equitable approaches to develop healthy and sustainable neighborhoods. Since 2014 it has teamed up with multiple federal, state, regional, and local partners to sponsor *Local Foods, Local Places* to support community-driven efforts to preserve open space and farmland, boost economic opportunities for local farmers and businesses, improve access to healthy local food, and promote childhood wellness.[13]

The following chapters cover a wide range of state and local policy options that can be used to advance resilient and sustainable food systems. They are not meant to be prescriptive but instead to lift up and share examples so people can adapt, borrow, and learn from each other.

Chapter 5 addresses land use planning with an emphasis on local policies, especially zoning. It highlights various kinds of zoning as well as things to consider in zoning ordinances. It also addresses subdivision and other ordinances, and a few broader approaches to land use that hold promise for food systems efforts. Chapter 6 focuses on policies to sustain farms and farmland. Recognizing the tension between new neighbors and agriculture, every state has done something to address competition for land, nuisance complaints, and farmland conversion. Most have programs to permanently protect farmland and to lease state-owned land for agriculture. Many support the economics of farming and ranching, legislate water rights, and address agriculture's environmental impacts. Chapter 7 covers food policies to overcome barriers to securing ingredients for a nutritious diet. While SNAP is the bedrock, many state and local governments support healthy retail policies to provide more nutritious food to residents who lack grocery stores close to where they live. They provide nutrition education and promotion and have programs to give away surplus food through food banks and other feeding sites. Chapter 8 takes a look at a few evolving issues, including farmland access, solar siting, and cannabis regulations, and offers some concluding thoughts.

Discussion Questions

1. What roles does the federal government play in food systems?
2. Which federal agencies might you engage in a food systems planning process?
3. Which nutrition assistance programs are likely to have the most impact on local and regional food systems?
4. How can conservation district staff help advance sustainable and resilient food systems?

Notes

1. *Federal Food Safety Oversight: Additional Actions Needed to Improve Planning and Collaboration* (Washington, DC: U.S. General Accountability Office, December 2014).
2. Vermont Law School Center for Agriculture and Food Systems and Harvard Food Law and Policy Clinic, Blueprint Project.

3. Congressional Research Service, "Federal Land Management Agencies: Background on Land and Resources Management," EveryCRSReport.com, February 9, 2009.
4. Richard Barrows and Lawrence W. Libby, "The Federal Role in Land Use Policy: Arguments For and Against Federal Involvement," 1982.
5. James Howard Kunstler, *Geography of Nowhere* (New York: Free Press, 1994).
6. Farmland Information Center, "Farmland Protection Policy Act Fact Sheet," *American Farmland Trust*, July 2022.
7. Renée Johnson and Jim Monke, *Farm Bill Primer: What Is the Farm Bill?* (IF12047, Version 4, Updated) (Washington, DC: Congressional Research Service, February 22, 2023).
8. Renée Johnson and Jim Monke, "Farm Bill Primer: What Is the Farm Bill? (IF12047, Version 4, Updated)."
9. USDA Food and Nutrition Service's Supplemental Nutrition Assistance Program, "Farm to School Census and Comprehensive Review," July 15, 2021.
10. *Federal Food Safety Oversight: Additional Actions Needed to Improve Planning and Collaboration* (Washington, DC: U.S. General Accountability Office, December 2014).
11. Donnie F. Williams, Ellie Falcone, and Brian Fugate, "Farming Down the Drain: Unintended Consequences of the Food Safety Modernization Act's Produce Rule on Small and Very Small Farms," *Business Horizons* 64, no. 3 (May 1, 2021): 361–68.
12. U.S. Department of Interior, Bureau of Land Management, "Livestock Grazing on Public Lands," accessed March 23, 2023.
13. U.S. Environmental Protection Agency, "Local Foods, Local Places," Overviews and Factsheets, June 6, 2014.

5
LAND USE POLICIES THAT SUPPORT FARMS, FARMLAND, AND FOOD

Chapter Summary

States grant local governments authority for land use planning, but some retain strong land use authority themselves. Local governments rely on policies like zoning and local ordinances to manage growth and development, and sometimes to support local farms, foster food production, protect the environment, and increase access to healthy food.

This chapter provides an overview of state and local land use policies. It describes different kinds of zoning and zoning considerations relevant to food systems, and other policies that can keep farmland in farming and support fair, sustainable, and resilient food systems.

***

As outlined in Chapter 2, most land use decisions are made at the local level. States grant local governments authority for land use planning, and local governments rely on policies like zoning and subdivision regulations to manage development and encourage—or discourage—agriculture and food systems activities.

At the state level, strong land use planning authority is a cost-effective way to balance community priorities such as housing and transportation with retaining land for food and other farm production. According to American Farmland Trust's Agricultural Land Protection Scorecard,[1] between 2001 and 2016, states with low planning scores converted five times as much land per new resident as states with high scores.

Eighteen states require communities—at least communities of a certain size—to develop comprehensive plans, but only 12 have land use goals related to compact growth or farmland protection. Programs that successfully support agriculture have three important features:

1. Goals to protect agricultural land and to promote compact growth;
2. Requirements to create comprehensive plans and adopt local policies to protect agricultural land; and
3. Consistency between state goals and local plans either through planning requirements or incentives to encourage conformance.

ALIGNING STATE AND LOCAL GOALS IN OREGON

Figure 5.1 Willamette Valley around 1970.
Source: Oregon State Parks Department.

Figure 5.2 Willamette Valley 2023.
Source: Oregon Department of Land Conservation and Development. *Google Earth.*

Oregon has a leading land use planning law that has guided development since being approved by voters in 1973. Among many achievements, it has retained more than 16 million acres of farmland while Oregon's population more than doubled. A key to its success was requiring counties to inventory farmland and adopt zoning and other policies to protect it.[2]

Table 5.1 Average Annual Percent Population Change: Oregon vs. U.S. 1970–2020.

	1950	1960	1970	1980	1990	2000	2010	2020
Oregon	39.6%	16.3%	2.26%	0.80%	1.98%	1.16%	1.02%	0.36%
U.S.	14.5%	18.5%	1.10%	0.95%	1.23%	0.95%	0.74%	0.25%
Difference	25.1%	-2.2%	1.16%	-0.15%	0.75%	0.21%	0.28%	0.11%

Source: U.S. Census Bureau, "Historical Population Change Data (1910–2020)," April 26, 2021.

After World War II, rapid population growth triggered Oregonians' concern about the future of agriculture and forestry and the impacts of sprawling development on the cost of community services. This led to the passage of the Land Conservation and Development Act, a powerful partnership between the state and local governments. It requires

> local governments to adopt comprehensive plans and state agencies to comply with them, sets standards to guide planning and development, and imposes penalties on communities that do not conform, including suspension of local revenue support.[3] It explicitly calls for farmland preservation and addresses other natural resources like water and forests, along with housing, transportation, and economic development.[4]
>
> Today, all Oregon counties have adopted comprehensive plans, representing years of effort to build consensus between citizens and local officials. Plans include background information on each of the state's 19 planning goals and a policy element describing the community's own long-term goals and implementation measures to achieve them. Implementation measures include urban growth boundaries to accommodate 20 years of projected growth, and effective farm use (EFU) zoning. EFU zoning requires minimum parcels of at least 80 acres for farmland and 160 acres for rangeland, although counties may adopt a lower minimum in some cases.[5] In exchange, landowners get lower property taxes and nuisance protections.
>
> Oregon's law has curbed land speculation and staved off land use conflicts by emphasizing coordination between local plans and statewide goals. To ensure consistency, communities submit a package of materials to the state, which includes proposed ordinances and other policy tools to achieve the statewide planning goals. Once approved, the package is acknowledged by the Land Conservation and Development Commission—a body comprising seven appointed volunteers from different parts of the state responsible for establishing the program's goals and policies.

Land use planning and zoning are related but often conflated. Land use planning is a structured process; zoning is a regulation. They are best done together but often are not. Much of this chapter focuses on zoning because it is so widely used. But communities also use subdivision regulations and other tools, so we will start there.

Subdivision Regulations

Subdivision regulations guide the division of land into lots, blocks, streets, and neighborhoods for the purposes of a legal sale. They establish standards to ensure adequate infrastructure for new developments, including utilities, road access, water, sewer, and so on. They usually require an overall map called a subdivision plat, which must be approved by a local planning authority. Once approved, it is recorded and becomes public record. Sometimes they require rezoning.

Zoning and subdivision regulations may be used at the same time. Both are governed by state law, but subdivision regulations tend to be more technical and less political than zoning. Best known for creating standards for housing developments, subdivision regulations also can address community connectivity by limiting or prohibiting dead-end streets or requiring road or sidewalk connections between adjoining subdivisions. They also are used to guide repurposing industrial sites; to encourage walkable communities, biking, and public transit; and to protect natural resources such as wetlands, floodplains, and open land. Some subdivision regulations address agriculture and conservation developments, including Agrihoods and Cluster Developments.

Agrihoods

Agrihoods—aka *Development Supported Agriculture*—are an innovation in master-planned communities (MPC). MPCs are large residential communities typically built by a single developer around a theme like senior living or recreational amenities like a golf course. In the case of Agrihoods, they are built around small, often organic farms.

A 2015 study found that 73 percent of U.S. residents consider access to fresh, healthy foods a high priority when deciding where to live.[6] Agrihoods respond to this demand by setting aside land for farming and protecting it with a conservation easement or other property covenant. Residents may farm land themselves, lease land for farming, and/or own shares in the farm so it operates like a CSA. The farm may include a U-Pick operation or other ways to engage residents, or residents may hire a farm manager to live on-site and manage the farm on their behalf.

Prairie Crossing was one of the first. It was created in the Chicago suburb of Grayslake, Illinois, in the mid-1990s to advance responsible development practices, conserve open land, and provide commuters easy access to rail services. It includes 100 acres of protected farmland used by the Prairie Crossing Farm, a nonprofit that supports multiple independent farm businesses, offers educational programming, and provides residents as well as the general public with produce, eggs, and other farm products.[7]

Cluster Development

Cluster development—aka *Conservation or Open Space Development*—requires houses to be grouped on small lots and protects the remaining open land for recreation, gardening, and occasionally small-scale agriculture. The protected land often is owned by the developer or a homeowner's association and sometimes leased to a farmer. In this case, it is most effective when open space requirements are mandatory and the open land is protected by a conservation easement.

Cluster developments work best for agriculture in transitional areas and when they create a buffer between residential development and working farms. Often used with zoning, they generally require updating ordinances and redefining frontage, lot size, setbacks, and other zoning regulations.

Transfer of Development Rights

Transfer of development rights (TDR) programs—aka *Transfer of Development Credits and Transferable Development Units*—pay landowners to limit development on some or all of their land. They combine planning and zoning to advance a community's goals for growth and development in tandem with resource protection goals.

TDR programs leverage the private marketplace to shift development away from farmland and other resources identified for protection (sending areas) to designated growth zones (receiving areas). Private developers buy development rights within the sending area in exchange for development incentives in the receiving area. Incentives include density or intensity bonuses, reduced parking requirements, and other regulatory flexibility.

Some programs allow developers to make monetary payments, leaving it to the local government to buy development rights, often in partnership with a public *Purchase of Agricultural Conservation Easement* (PACE)—aka *Purchase of Development Rights* program—(see Chapter 6) and/or a local land trust. Others buy and retire the rights to stimulate the market and/or reduce overall building potential. Still others establish TDR banks to buy development rights with public funds and then sell them to developers.

Tying TDR to zoning gives local governments control over where and how growth will occur and reduces the windfalls and wipeouts in property values often associated with zoning changes. A well-known example is Montgomery County, Maryland, which has protected over 50,000 acres with TDR and over 20,000 acres using county funds for PACE. Combined, its efforts have resulted in the highest percentage of farmland protected by agricultural conservation easements of any county in the country.[8]

TDR IN CHESTERFIELD TOWNSHIP, NEW JERSEY

In southern New Jersey, Chesterfield Township has a population of about 7,500. It combined TDR with state and county PACE funding to protect nearly 8,000 acres—most of its farmland and over half of its total land area.

In 1998, Chesterfield adopted a new Land Development Ordinance that enabled TDR, downzoned farmland in agricultural zones to 10-acre lots, and added a clustering option to give developers a density bonus in exchange for protecting 50 percent of the property as open space. Launching the program involved extensive public engagement and active participation from the development community (see Figure 5.3).

Based on community input, the township created a new neighborhood called Old York as a receiving area. Today Old York is a walkable, traditional community with density comparable to its historic neighbor, the village of Crosswicks. Development allowances are graduated based on the type of housing, and 6 percent of its units are low–moderate-income housing.[9] Its sending areas contain nearly 7,500 acres of farmland. TDR credits are based on soil quality and on-site septic systems and allocated using a formula that considers preexisting zoning and environmental constraints, using computerized soils data to determine how many TDRs to allow on each parcel. Future development on protected farms is limited to one dwelling per 50 acres.

The county's Development Credit Bank facilitates transfers by buying TDRs at competitive prices and selling them when there is unmet demand from the private marketplace. As of 2022, the program has preserved 46 farms on 2,772 acres, along with another 48 farms on 5,043 acres, through a variety of traditional PACE programs.[10]

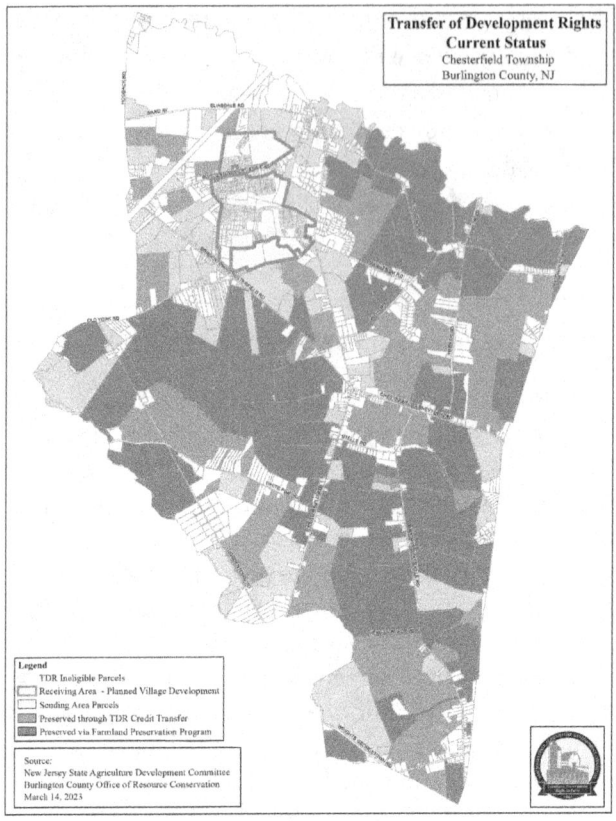

Figure 5.3 Chesterfield Township TDR receiving/sending areas and farmland properties permanently protected with TDR and PDR.
Source: Map courtesy of the New Jersey State Agriculture Development Committee.

Urban Growth Boundaries

Urban growth boundaries (UGBs) or urban service boundaries promote efficient and orderly land use and development. Their purpose is to identify areas to encourage urban development and areas to protect and support rural land uses. Higher density and mixed-use developments take place inside the boundary, along with supporting amenities like parks and a full suite of public services provided. Agriculture and other low-density land uses occur outside the boundary and receive limited public services.

UGBs are designated on land use maps and used to guide decisions on infrastructure and other development, especially the extension of roads and services like public water and sewer. They are set for a specified time period—usually 20 years—to accommodate projected population growth and land use needs. They can be renewed at the end of the period or adjusted if conditions change. Oregon's UGBs are well known and often studied, but the first was enacted in Lexington, Kentucky (see box).

GROWING IN NOT OUT IN KENTUCKY

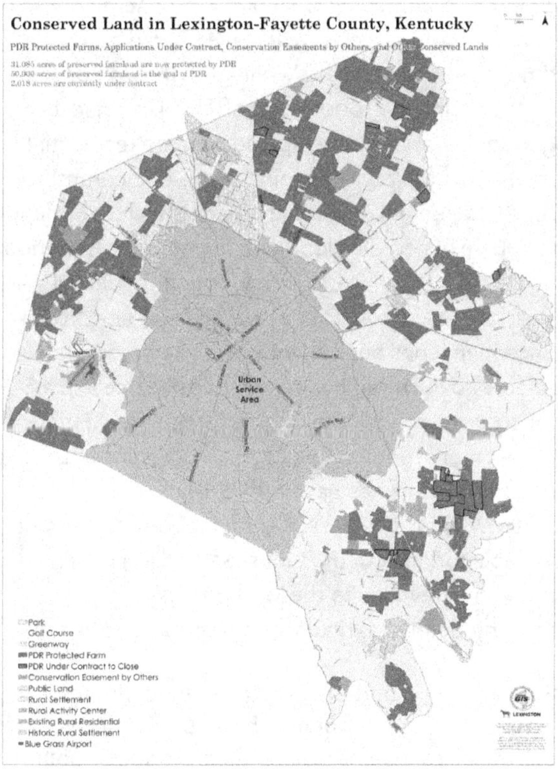

Figure 5.4 For over 60 years, Lexington, Kentucky, has grown sustainably through compact development within its USB, while Fayette County has preserved farmland, natural resources, and rural character.
Source: Map courtesy of the Lexington-Fayette Urban County Government GIS Section.

Fayette County is part of the Bluegrass Region, considered the horse capital of the world. Famous for rolling hills and deep residual soils, its farmland is rich in calcium and phosphorous, which naturally fortify plants and animals, including strong bones for horses. It is home to the city of Lexington and the nation's first urban service boundary (USB), created in 1958 to control sprawl and limit the costs of community services. Nearly a third of its land base and about 97 percent of the population lives within the USB.

In 1964 the city established a 10-acre minimum lot size in the rural district. A decade later, the city and county created a merged government, which led to a unified approach to land use planning and coordinated municipal/county actions to achieve shared goals.

It took time to get things right. While growth was concentrated within the USB, by 1997, almost 5,000 acres of farmland had been developed for housing in the rural area. The spread of large residential lots led Lexington's mayor, Pam Miller, to appoint an ad hoc committee to develop a Rural Land Management Plan. They made two recommendations that changed the course of new development: Increase the minimum lot size to 40 acres and create a Purchase of Development Rights (PDR—aka PACE) program. In 2000, Lexington/Fayette passed the first PDR legislation in the state. Since then it has permanently protected more than 30,000 acres of its 50,000-acre goal. The ad hoc committee became a quasi-governmental, nonprofit corporation, Fayette County Rural Land Management Board, Inc., which now purchases and holds agricultural conservation easements. This integrated approach has saved infrastructure costs, enhanced connectivity, created vibrant businesses, and supported a vibrant agricultural sector responsible for one out of 12 jobs, $2.3 billion in annual output, and about $8.5 million of revenue to the local tax base.[11]

Current zoning defines agricultural use broadly to mean:

> the use of a tract of land of at least five (5) contiguous acres for the production of agricultural or horticultural crops, including, but not limited to, livestock; livestock products; poultry; poultry products; grain; hay; pastures; soybeans; tobacco; timber; orchard fruits; vegetables; flowers or ornamental plants; including provision for dwellings for persons and their families who are engaged in the above agricultural use on the tract, but not including residential building development for sale or lease to the public.[12]

> It defines urban agriculture within the USB and includes community gardens and agritourism so long as they are subordinate to and directly associated with principal agricultural production. The ordinance also includes an Agricultural Market Overlay Zone, which allows sales facilities, racetracks, stables, and accessory uses with large facilities often at least 40 acres in size. Agricultural markets are designated exclusively for the purpose of buying and selling farm products, including a stockyard, and include any products grown, raised, or made by agricultural producers[13] (see Figure 5.4).
>
> Local leaders continue to innovate. In 2014, the office of Economic Development created Bluegrass Farm to Table, which convenes a Local Food Systems Stakeholder Group and facilitates coordination and collaboration within the local food economy. It maintains strategic partnerships and supports agricultural economic development from farmers markets and farm-to-school to aggregation, distribution, and processing infrastructure for local farm and food products. Key stakeholders include farmers, processors, distributors, buyers, community groups, and policymakers. Further, in 2022, the county Health Department began offering CSA vouchers to employees interested in signing up for farm shares.

Zoning

Zoning regulates the what, where, and how of different land uses. Based on the purpose and power granted in 1926 by the Standard State Zoning Enabling Act, it divides jurisdictions into districts and sets standards for development, including lot sizes, setbacks, signage, and parking. Enacted to protect health, safety, and general welfare, it is a powerful tool but can be preempted if it conflicts with state or federal law.

There are various forms of zoning. Mostly used to guide growth and development, they also can protect agriculture and promote food production, marketing, and retailing.

- *Euclidean zoning* is the most familiar and traditional form. It is named for the 1926 Supreme Court decision *Village of Euclid, Ohio v. Ambler Realty Co.*, which upheld zoning codes as a valid extension of a city's right to regulate land uses. It separates incompatible uses by dividing them into broad categories, typically residential, commercial, and industrial devel-

opment, often also including institutional land uses and rural, agricultural, and/or open space.
- *Form-based zoning* is a modern alternative to Euclidean zoning. Instead of focusing on land use, it addresses the physical forms and relationships of buildings to streets, sidewalks, and other aspects of the public realm.
- *Incentive zoning* provides a reward system to encourage certain kinds of development rather than relying on regulations.
- *Intensity zoning* is based on how much density is allowed, such as number of residential or commercial units.
- *Mixed-use zoning* allows multiple uses within one district that otherwise would be confined to separate zones. This can strengthen community food systems by supporting local food production while increasing access to grocery stores and supermarkets.
- *Performance zoning* is market oriented and uses effects-based criteria to guide proposed developments. It addresses private property rights, especially in terms of environmental protections.

Zoning ordinances define the scale, scope, and intensity of allowable uses. They include a definition of terms, maps, and text that outlines their authority and intent, jurisdictional reach, uses and restrictions, development standards, and administrative procedures. When defining terms like "farm," "food production," and "healthy food retail," proactive ordinances use broad definitions to include activities like farmers markets, retail sales, urban agriculture, and enterprises that process and add value to raw agricultural products. To increase local food production, consider allowances for season-extending facilities, such as greenhouses, hoop houses, and high tunnels. To promote healthy food retail, consider exemptions and incentives to encourage siting stores in underserved communities (see Chapter 7).

Ideally, zoning is tied to the vision and goals of comprehensive and other relevant plans. Implementation relies on community support. Zoning can be a dirty word if community members believe it infringes on their property rights rather than protecting their quality of life.

Agricultural Zones

Many suburban and rural communities use rural-residential, or rural-agricultural zones, which allow but do not prioritize agriculture. Except sometimes by name, this is not agricultural zoning. In fact, the resulting large-lot, low-density residential development consumes more acres for fewer, often larger houses. This tends to fragment the agricultural land base and inflate property values beyond what most producers can afford.

Between 2001 and 2016, agricultural land in low-density residential areas was 23 times more likely to urbanize than other agricultural land, driving farmland conversion in the short term and paving the way to more urbanized development.[14] Large houses on dispersed lots have an outsized impact on climate change, requiring more energy to heat and cool than other residential developments and leading residents to drive more because they have less access to services. A study of 93 million U.S. households found that due to larger homes, wealthier Americans have per capita footprints ~25 percent higher than those with lower incomes, and emissions from affluent suburbs can be 15 times higher than nearby neighborhoods with smaller homes.[15]

Agricultural zoning addresses these issues by designating farming and/or ranching as the preferred land use. It limits activities that do not support agriculture and sets performance standards for accessory and ancillary uses. Some agricultural zones limit activities to those conducted by farm families or employees, or to uses that will not impede transfer to a bona fide farmer. Purpose, more than lot size, makes the difference. With strong nuisance protections, vegetative buffers, setbacks, and other farm-friendly ordinances (see Chapter 6), large-lot residential zones can be retooled to support agriculture and food production.

Agricultural Protection Zones

Agricultural Protection Zones (APZ) stabilize the land base by directing new development toward community facilities and away from farms and ranches. While each APZ states its own intent, their overall purpose is to support the agricultural economy and ensure that important agricultural soils remain available for farming and/or ranching. Often, APZs have the added benefits of fending off nuisance complaints and maintaining rural character.

Zoning that limits density may reduce property values, while zoning that limits farm labor housing or the size of farm structures may restrict operations so much they no longer are viable. The best way to avoid unintended consequences is to engage the agricultural community before imposing new regulations and to allow flexibility to modify ordinances to support new approaches to agriculture and food production.

APZs take several forms. *Fixed density*, often based on existing land use, limits non-farm development, for example, allowing one dwelling for every 25, 40, 80 or more acres. The challenge is determining the appropriate lot size for different kinds of agriculture and ensuring that lots do not get turned into estates that drive up land values and threaten agricultural viability. *Area-based allocations* are a creative solution to these problems. Allowing some residential development while achieving agricultural protection, allocations may be based on acreage or percentages of the total parcel—for example, a 90/10 ratio of farmland to development. *Sliding-scale zoning* is a common form of area-based allocation. It is based on farm size to provide more flexibility on new buildings and acreage allowances, so the larger the parcel, the greater the number of buildings allowed. For example, a 40-acre parcel might be allowed one or two dwellings, while a 400-acre parcel might be allowed ten. To retain as much farmland as possible, ordinances often adjust road frontage requirements, require clustering to achieve a specific ratio, or tie to soil quality.

Effective or exclusive agricultural zoning limits land uses to farming and/or ranching and to operations that support it. For example, in Wisconsin, lands within an exclusive agricultural zone must be within a county agricultural preservation plan and the ordinance must both be consistent with the plan and—to qualify for tax credits—certified by the Wisconsin Land and Water Conservation Board. All structures and conditional uses within the district must be consistent with defined agricultural uses so that the zoning cannot be challenged in court as exclusionary. *Intensive agricultural zones* go further to separate residential development likely to conflict with agricultural operations and restrict dwellings to those related to agriculture. They are appropriate in places with intensive agriculture uses such as vineyards or CAFOs, especially when combined with buffers and/or setbacks. In areas where the zone supports local food farms and specialty crops, smaller lot sizes may be allowed along with commercial and industrial uses that are compatible with and enhance agriculture.

ZONING FOR AGRICULTURE IS IN SCOTT COUNTY'S BLOODSTREAM

Figure 5.5 Corn for grain is Scott County, Iowa's leading crop.
Source: USDA. Photo by Preston Keres.

With strong political support and policy enforcement, agricultural preservation zoning has secured a land base for farming in Scott County, Iowa's third-most-populous county. Located 165 miles west of Chicago, it hosts the global agricultural equipment company John Deere and other major agribusiness employers, including Tyson Fresh Meats and Oscar Mayer/Kraft.

Settled before Iowa became a state in 1846, agriculture has always been important to Scott County's culture and economy. Its soils are some of the most fertile in the country. Despite strong development pressure, it has 220,000 acres of farmland in production, mostly in corn and soybeans, but also livestock and poultry, led by hogs and pigs. Average net cash farm income per farm is healthy at $176,328,[16] and the county ranks tenth in the state for nursery, greenhouse, floriculture, and sod.[17]

The county has used strong land use policies to support agriculture since 1980 when it created an Agricultural Preservation (A-P) zoning district and strong subdivision policies. The A-P zone furthers the goals of the county's comprehensive plan to protect highly-quality soils and agricultural operations from urban development.[18] Inclusion is based on soil quality. Metrics are determined by prime farmland as defined by NRCS's soil surveys and a Corn Suitability Rating (CSR) of sixty (60) or greater as a weighted average per quarter section of land. On farmland with a CSR of 60 or higher, only construction of farm-related housing and buildings is allowed. Once a Plat of Survey is recorded, no land or lots may be subdivided into smaller lots without rezoning, and rezoning requires county planning staff and NRCS to conduct in-depth soils studies. This carries over to subdivision applications to split existing housing from a farm.[19]

The A-P district allows farm processing. For example, County Supervisors allowed a farmer with 2,000 acres of row crops and more than 2,000 cattle to install a digester to convert manure and cover crop biomass into electricity and digestate fertilizer. Working together, these increase organic matter and improve soil health, generating both economic and environmental value.

The county also has other zones for agriculture. Its less restrictive Agricultural-General (A-G) district acts as a holding zone until a compatible development proposal is approved through special use permits or rezoning. It allows a limited number of public and private uses, such as churches and schools, and some commercial or industrial uses through overlay districts. It also has an Agriculture Commercial Service Floating (ACS-F) district to serve the farm community. This zone allows farm-related businesses and services to locate in certain unincorporated areas. Permitted uses include livestock auctions and large animal veterinarians; feed mixing and blending; grain handling; retail sales of seeds and inputs; and logistics businesses involving local transportation of agricultural inputs, commodities, and livestock.

In 2022, the county approved a solar ordinance to encourage renewable energy production but to limit utility-scale solar development to lower-quality soils. Individuals and farms can generate solar for their own use, but only utility-scale solar installations are allowed on soils below the CSR-60 threshold. The county also allows wind power but has not had much demand.

Elected officials have supported strong land use policies for decades. They understand their history and the importance of agriculture to the county. The community's values for protecting their valuable soils are in their bloodstream.[20]

Agricultural Overlay Zones

Overlay zones are a flexible tool used to modify zoning by adding new standards or regulations to accomplish specific goals and/or encourage certain kinds of development. Housing overlay zones are used to incentivize developers to build affordable housing by increasing density bonuses, reducing parking requirements, or streamlining permits. Overlays also are used to protect sensitive resources like aquifers and flood plains by limiting or even prohibiting development to protect water quality and groundwater recharge, and/or to provide floodwater storage capacity.

Agricultural overlay zones apply this popular tool to reduce conflicts between farmers and non-farm neighbors and/or to identify priority areas to support farming by waiving, instituting, and/or strengthening zoning provisions without having to redraft entire ordinances. Typically designed to protect quality soils and contiguous areas of working farms or ranches, they support the agricultural economy and maintain farmland affordability. They also can be used as a TDR sending zone to remove barriers for urban agriculture in districts where it is not expressly allowed or to support the economics of agriculture by supporting compatible commercial, marketing, and light industrial uses. These generally allow the same uses as agricultural zones while encouraging support industries like auctions, stockyards, and farm equipment rental and sales.

Urban Agriculture Zones

Zoning for agriculture in cities is different from zoning in rural areas. Urban agriculture often involves public land, occurs on small parcels, and may include backyards, vacant lots, rooftops, and indoor settings. California passed legislation in 2013 to create urban agriculture incentive zones (see Chapter 8), but most cities do not zone for it, thus existing codes may impede food production and distribution.

To encourage urban agriculture, local governments can follow California's lead and create incentive zones or overlay zones and special districts for urban farms and gardens. They also can include enabling language to indicate where and what kinds of agricultural activities are allowed in different zones. Considerations are similar to other kinds of agricultural zoning and extend to charitable, educational, and other non-commercial activities.

Figure 5.6 Taqwa Community Farm in the Highbridge neighborhood of the Bronx, New York City.
Source: USDA/FPAC. Photo by Preston Keres.

ZONING FOR URBAN AG IN TWO CITIES

Cleveland, Ohio, and Somerville, Massachusetts, both address urban agriculture through zoning. Cleveland allows community gardens, beekeeping, livestock, and greenhouses as a conditional use in all zoning districts, subject to certain regulations. It also created a special Urban Garden District that allows agriculture, community, and market gardens by right. The zones allow the sale of crops produced on-site, along with a series of permitted accessory uses: season-extending structures; signage; raised beds, compost bins, seasonal farm stands, fences, chicken coops, and beehives; and buildings including tool sheds, barns, and planting preparation houses.[21]

Somerville amended its zoning bylaws to add an Urban Agriculture Ordinance that allows hens, honey bees, and yard farms, including season-extending accessory structures. It also permits commercial farms and greenhouses, including hydroponic and aquaponic farms, on municipal land, buildings, and rooftops. Accessory uses are subject to building code requirements and public health regulations. Products from

> commercial farms may be sold on-site subject to a series of provisions. Two of these include an allowance for farmstands and proof of annual soil testing, which must be posted when produce sales occur.[22]

Zoning Considerations

Accessory and Ancillary Uses of Agricultural Land

Accessory uses serve but are subordinate to a zone's principal permitted uses. Examples include greenhouses and farm stands. Ancillary uses are secondary uses that should not interfere with the farming operation. Examples include renting office space in an outbuilding or leasing land to hunters.

Zoning ordinances define whether uses like agritourism, wine tasting rooms, and horse stables are agricultural, accessory, or ancillary. They spell out when uses are conditional and require special permitting. They often address farm-related services, such as equipment repair or commercial composting, and whether to permit non-farm activities. Once these are defined, ordinances usually include performance standards to certify uses are of a nature, intensity, scope, size, and appearance to conform to existing structures, and require they will not hinder future farming activities or limit the amount of land and/or structures allowed for specific purposes.

Eagle Mountain, Utah's zoning code defines agriculture to include row crops, livestock boarding, grazing, and associated activities, but does not allow CAFOs or animal by-product businesses. Accessory uses and structures are allowed if they do not substantially alter the character of the principal uses of structures and facilities like barns, greenhouses, and silos. Farm stands are allowed, but only to market products grown on the farm, and only if structures are 300 square feet or less. Conditional uses include value-added processing and equestrian facilities. Conditional ancillary uses include religious or cultural meeting halls, transmission towers, commercial hunting, and home businesses and accessory dwellings.[23]

Agritourism

Agritourism (aka *agrotourism* or *agritainment*) combines traditional agricultural activities with on-farm entertainment and/or education. Usually pursued to diversify farm income and stimulate rural tourism; its revenue more than tripled between 2002 and 2017, reaching nearly $950 million, not including wineries.[24]

Common forms include U-Pick activities where customers harvest various types of farm products themselves, for example, apples, berries, Christmas trees, and pumpkins. Some operations offer corn mazes, hayrides, and petting zoos. Others offer on-farm demonstrations, rent farm facilities for weddings and festivals, or operate farm-stay bed-and-breakfasts or dude ranches. Thus, agritourism is not always considered agriculture.

Most states have enacted statutes to address agritourism issues. These include zoning definitions and standards for activities allowed by right and other requirements. Some list agritourism as an accessory or ancillary use, with conditions and limitations. Others create stand-alone ordinances or overlay zones, often limited to authentic accessory uses that support and promote working farms and ranches rather than supplant them.[25] Issues to address include minimum parcel and maximum facility sizes, setbacks, parking allowances, signage, noise, nuisances, and hours of operation. In cases where heavy uses are allowed, like festivals, they address traffic control and neighborhood impacts as well.

Buffers and Setbacks

Setbacks and buffers separate land uses within zones. They are a useful tool to create space between farmers and non-farm neighbors to stave off conflicts, reduce trespass and vandalism, and protect natural resources.

Setbacks refer to the distance between a building and a property line, a natural resource like a wetland, or a road or other area that needs protection from new development. *Buffers* take many forms but usually are vegetative and designed to shield neighbors from objectionable smells, noise, dust, sounds, and so on. Some zoning ordinances require a minimum width, while others vary the requirement based on the type of buffer.

Buffers also are an important tool to protect soils, waterways, and sensitive habitats from agricultural runoff. *Conservation buffers* include wildlife corridors, greenways, windbreaks, and filter strips. *Riparian buffers* are adjacent to waterways and contain a combination of perennial plants, shrubs, and/or trees. They also can be managed to produce a harvestable crop.

Local ordinances take many approaches to setbacks and buffers. They may require one, the other, or both between existing farms and abutting new development. They may limit requirements to certain types of farms or development. Some combine them and require a buffer within a setback, while others simply require a minimum buffer distance.

Butte County, California, requires clear delineations for all lands zoned for agriculture, performance standards and land use transitions, setbacks, and buffers between residential development and agricultural uses. To protect farms, it requires 300-foot boundaries for bordering zones where boundaries abut agricultural zones.[26] Clark County, Washington, takes a more nuanced approach. It requires new buildings to comply with setbacks that range from 100 feet for livestock grazing or pasture to 300 feet for row crops and vegetables, with setbacks for other types of agriculture in between. It allows use of 8-foot berms to satisfy part of the setback requirement so long as they are contoured at three to one (3:1) slopes and include shrubs, trees, or grasses and achieve a finished height of 15 feet. Vegetative screens also may be planted continuously along the parcel line. Trees must be at least six feet high when planted and reach an ultimate height of at least 15 feet.[27]

Composting

Compost is created by combining organic wastes into piles, rows, or vessels. It can include leaves and other yard waste, uneaten food, manures, and bulking agents such as wood chips. It has many benefits:

- Compost can be added to soil to improve its fertility, promoting higher yields, and limiting petrochemical inputs;
- It provides carbon sequestration and improves water retention in soils;
- It reduces methane emissions by keeping organic waste out of landfills; and
- It improves contaminated and otherwise compromised soils, which aids reforestation, wetlands restoration, and habitat revitalization.[28]

Composting occurs on farms and in backyards as well as commercially. Zoning can address both scales, for example, by allowing it by right for farm-generated waste and through permit requirements for businesses. Most ordinances limit what types of wastes can be composted and have policies to mitigate odors and pests. They also set size allowances. Rural communities

Figure 5.7 Trent Deans, Veteran Compost D.C. works organic composting operations at the Arcadia Center for Sustainable Food and Agriculture in Virginia. *Source:* USDA. Photo by Tom Witham.

are likely to have more flexible policies than urban, like allowing farmers to compost municipal leaves and other off-site materials.

Commercial composting generally is regulated by state as well as local governments. The U.S. Composting Council tracks legislation and created a Model Compost Rule Template for states to adopt.[29] Commercial facilities have larger impacts and generally require plan approval and setbacks, especially if in or near residential zones. Local governments also may consider regulations to prevent stormwater runoff, screen large sites from public view, and address neighborhood issues like dust, pests, and odors.

Farm Labor Housing

Safe and suitable lodging for farm labor is important, especially for dairy and specialty crop operations. Zoning ordinances vary widely but generally are fairly flexible so long as standards comply with public health and safety laws. They may allow mobile homes, cabins, or other temporary housing for seasonal workers, or allow building a second or third house without triggering subdivision regulations.

Burlington County, New Jersey, has a model farm labor housing ordinance for municipalities to use to support sectors like equine, nursery, and local food production, which require significant and/or specialized labor. For permanent structures, it considers impacts on farm operations and the size, number, and type of workers to be housed. It requires parcels to be at least ten contiguous acres and for occupants to use the same driveway as the farm operation or principal residence. Seasonal housing is allowed as a conditional use, subject to setbacks, buffers, and additional requirements.[30]

Healthy Food Retail

Zoning can be used to increase the availability of healthy food options and/or to limit fast food establishments, especially near schools. New York City uses zoning exemptions in its Food Retail Expansion to Support Health (FRESH) program. Created in 2009, FRESH was a response to a citywide study that highlighted a shortage of fresh food options in several neighborhoods. It offers a suite of zoning and financial incentives to support stores in underserved communities that provide a full range of grocery products, including fresh meat, fruit, and vegetables. Zoning incentives include larger building allowances and reduced parking requirements in mixed residential and commercial districts.[31]

Local governments also use zoning to limit fast food outlets. Often enacted for aesthetic reasons, some aim to improve health outcomes.[32] Cities like Arden Hills, Minnesota, and Detroit, Michigan, have regulations requiring at least 500 feet between fast food outlets and schools.[33] Others regulate the density of fast food restaurants, while others restrict the development of fast food restaurants altogether (see Chapter 7 for other healthy retail policies).

Local Food

In June 2022, USDA announced a framework to shore up supply chains to transform food systems to be more resilient, competitive, and fair. It called for more regional processing and for food systems to become more distributed and localized. Achieving this will depend, at least in part, on zoning and other local land use regulations.

Zoning ordinances are rarely explicit about local food but still address it in several ways. Approaches vary, especially between urban and rural communities, but both can use zoning to expand production, aggregation, processing, marketing, and distribution.[34] In rural and peri-urban areas, this could mean adjusting ordinances to support smaller acreage farms, like those producing fruits and vegetables and supporting value-added processing, on-farm marketing, and associated activities from cooling to composting.

On-Farm Marketing

Zoning ordinances often allow local food sales through farm stores, stands, and other forms of on-farm direct marketing by right or as an accessory use. Either way, it is important to determine whether this is a broad right or limited to seasonal produce.

Farm stands and stores range from seasonal and temporary, like a covered wagon with a cash box, to year-round, full-service stores that sell farm-themed

Figure 5.8 Tracy Potter-Fins and Bethany Stanbery grow fresh, high-quality, certified organic, Montana homegrown produce and flowers for their community through their farm stand and other markets.
Source: USDA/FPAC. Photo by Preston Keres.

items along with farm-grown products. Effective farmstand ordinances have clear definitions to distinguish between the scale and scope of different types of operations. A small, seasonal farm stand does not require the same development standards as a store with a commercial kitchen that serves food and has bathrooms and parking facilities. Beyond scale, scope, and seasonality, zoning considerations include market permanence, along with issues such as road setbacks, parking, lighting, and signage. Ordinances often stipulate that most—if not all—of the products must be grown on the farm where they are sold, or that 50 percent or more of what is sold must be grown in the state or region. They also may limit, or even prohibit, the sale of products like garden gnomes and other decorations.

Community Supported Agriculture (CSA) farms are supported by selling shares in a farm's yield for the length of a growing season. In exchange, members typically receive a weekly distribution of seasonal products. Some zoning regulations allow CSAs in residential neighborhoods, while others confine them to agricultural zones. Zoning may address permitted and accessory uses, as well as parking and infrastructure requirements to handle weekly pick up of farm shares.

Farmers' Markets

Farmers' markets are community places where farmers come together to sell their products directly to consumers. Supporting farm viability and community food security, they attract foot traffic to downtown areas and increase social cohesion. Usually located in or near a large town or city, they often are managed by a government agency, civic organization, or other local entity. Depending on their location, size, and frequency, they may take place seasonally in a facility like a park or parking lot, or year-round in a permanent structure.

Zoning can designate where farmers' markets operate and the conditions under which they are allowed. Some communities have a straightforward permitting process and allow farmers' markets by right across all or multiple zones. Others require a special permit and restrict them to commercial, mixed-use, or other specified zones. Ordinances typically address when and how the market may operate, which products may be sold, signage, parking, and transportation issues. Some require 100 percent of the products sold to be grown by the farm that is selling them, while others only require 50 percent. Many allow music and other entertainment, and some allow the sale of local crafts.

ZONING TO PROMOTE FARMERS' MARKETS IN FRESNO, CALIFORNIA

Figure 5.9 Moua Farms sells farm-fresh vegetables to folks at the Brentwood Farmer's Market in Fresno, California.
Source: Eddie Ledesma/Contra Costa Times/ZUMA Press, Inc./Alamy Stock Photo.

Fresno County, California, is one of the nation's top agricultural counties. Supplying national and global markets, it produces hundreds of crops, including fruits, nuts, vegetables, dairy, poultry, and livestock. Still, until 2008, it was hard for people living in Fresno to find fresh, local food.

Fresno has very high rates of poverty and food insecurity, especially among minority households, including its farming populations. Health advocates became concerned about the relationship between food insecurity and rising rates of diet-related diseases. Working with the city of Fresno's planning department, community members and farmers, they gained unanimous support from City Council to change the zoning code to recognize farmers' markets.

Today, Fresno's Citywide Development Code has a full set of standards to guide where farmers' markets may be located, developed, and operated. It allows several types of markets. They range from permanent indoor markets that are regulated as Healthy Food Grocers to temporary farmers' markets in off-street locations using portable structures. It requires unprocessed farm products to occupy at least 60 percent of retail space, but products like bread and cheese are allowed—and beer and wine if vendors are licensed and have obtained required approvals from other city and state departments. Site plans must show the location of market stands, restrooms, and parking and require a traffic control plan showing street closures, detour routes, barricade locations, and traffic control signage. Other requirements address operation times, insurance, public notice, site development standards, layout, and signage.[35]

Mobile markets are essentially farmers' markets on wheels, where fresh local produce is transported in a bus, van, or other large vehicle to communities with limited fresh food options. Models vary widely: some represent a single operation, others aggregate produce from multiple farms, and still others procure from wholesalers, produce auctions, and food donations. Often subject to farmers' market and parking lot regulations, some are allowed on private property by right, while others are considered conditional use or require a special permit. Some ordinances dictate the size or grade of allowable vehicles, but usually they are allowed in parking lots of facilities with public access, like places of worship, community centers, and schools. Some require that a percent of the foods sold are WIC-approved.

Incubators and Shared Facilities

Finally, zoning can support community kitchens and incubators, which allow for shared use of food processing equipment to add value to local food and farm products. Some ordinances allow these uses by right in all zoning districts, others limit them to commercial or industrial zones. Food innovation districts are a novel approach. They include a range of services from food business incubators and shared facilities to food hubs, which connect producers with institutional buyers. Traverse City, Michigan, developed one on a former hospital campus. It includes a commercial kitchen, cold storage, a food hub, processing facilities, and a year-round farmers' market, along with housing, restaurants, and retail outlets. It has several zoning designations, including brownfields, historic district, and a Michigan tax-free Renaissance Zone.[36]

On-Farm Energy Production

Wind turbines, solar arrays, and methane digesters are valuable sources of renewable energy to support agricultural operations and can be sold to utilities for additional income. These may use or adapt existing land and structures without competing with the primary agricultural purpose. Utility-scale energy production, especially solar, is a more complex issue that generally competes with agriculture and food production. Thus, zoning often allows energy production for agricultural use by right but is more restrictive about commercial installations.

Strong ordinances are clear about these definitions. They include both permitted and conditional uses, where and under what circumstances renewable energy production is appropriate on agricultural land and in agricultural zones, and address setbacks and visibility issues. They can require

Figure 5.10 A free stall heifer barn with 720 photovoltaic solar panels at Brubaker Farms in Mount Joy, Pennsylvania, makes it both a dairy and green energy producer. The farm also has a methane digester, which can handle more than 41,859 metric tons of organic waste and which fuels a low-emission generator producing 225 kW that powers the digester and farm operations. Excess power is sold to the local power grid, allowing the community to benefit from a green energy source.

Source: USDA. Photo by Lance Cheung.

utility-scale facilities to be planned, developed, and decommissioned to limit adverse impacts on agricultural soils and to encourage continued agricultural use. Further, they can direct these projects to marginal and disturbed lands, like parking lots, brownfields, and landfills, and encourage developers to create and maintain pollinator habitats by planting flowers and native grasses around facilities, which also serves to reduce erosion and improve soil health. (See Chapter 8 for a deeper discussion of solar siting.)

Manure contains organic matter and useful nutrients, including nitrogen and phosphorus, which plants need to grow. When properly managed, it can be recycled and used as a fertilizer, soil amendments, and renewable energy. Anaerobic digester systems convert manure into biogas to generate electricity. Sometimes digesters are considered part of a farming operation and allowed in an agricultural zone, as in Scott County, Iowa. Other times they are considered a conditional or accessory use with strict guidelines. Lowville, New York, prohibits large anaerobic digesters but allows small digester systems as an accessory use on lots of ten acres or more in any zone that allows agriculture. It requires a permit, 100-foot setbacks from side and rear property lines and public road rights-of-way, and 300-foot setbacks from any residential structure other than that of the property owner.[37]

Poultry and Livestock

Most states have livestock regulations that supersede local zoning. Still, many communities have developed additional ordinances to address local nuisance, environment, and animal welfare issues. These vary widely between urban and rural areas based on type of operation and scale, but often have guidelines for site suitability, buffers and setbacks, and generally accepted agricultural (GAP) practices. Key considerations include the number of animals allowed per acre based on conditions like soil quality, water availability, and the space needed to provide shelter and avoid crowding.

Concentrated Animal Feeding Operations

CAFOs are regulated by federal laws such as the Clean Water Act and state laws, including Right to Farm Laws (see Chapter 6). Local governments use zoning to further regulate animal waste, which can impair water bodies, contaminate drinking water, produce GHG, and smell bad. Some communities use siting and permitting standards to guide where and to what extent CAFOs are allowed. Others create moratoriums to study their impacts, and some simply prohibit them.

Beyond siting requirements, some ordinances require operators to post bond, while others empower local boards of health to monitor CAFOs and create policies to protect public health. Ordinances that address siting typically use density requirements, odor controls, buffers, and setbacks to limit negative impacts. Cerro Gordo County, Iowa, requires siting CAFOs and commercial feedlots at least one-quarter mile from residential and commercial districts, and storage lagoons at least 200 feet from a property line or road right-of-way.[38]

CAFOs also are associated with animal cruelty due to close confinement and crowding, which create boredom and stress in the animals and lead to physical and mental illnesses. However, these issues largely have been addressed through state ballot measures rather than zoning.

PERMITTING CAFOS IN BAYFIELD, WISCONSIN

Bayfield County, Wisconsin, developed a careful permitting process for CAFOs along with licensing and setback requirements. They are based on but exceed the state Department of Natural Resources' regulations. Intended primarily to protect water quality, the state does not regulate facility siting, odor control, hazardous emissions, or animal welfare. So Bayfield's ordinance spells out a series of conditions to protect human and animal health, prevent pollution and nuisances, and preserve the quality of life, environment, and small-scale livestock and other farm operations.

Bayfield's ordinance uses the state's definition of livestock operations with 1,000 animal units or more. It includes provisions for two similar smaller operations if they operate together. Operations meeting this threshold go through an extensive permitting and approval process, including a public hearing. Applicants also must file a financial assurance in case they need to clean up environmental contamination, and/or abate public nuisances caused by operations—including but not limited to testing and replacement of contaminated wells and water supplies—and to ensure proper closure if for any reason the applicant decides to shut down. It also requires a nutrient management plan, and while largely focused on environmental quality, it also addresses animal and employee welfare and emergency management.[39]

Urban Chickens, Bees, and Livestock

Most communities regulate poultry and livestock through animal codes instead of through zoning, but cities and towns increasingly are amending zoning codes to allow poultry, small livestock, and honey bees. Usually their goal is to strengthen community food security by increasing a local supply of eggs and honey, and sometimes meat or milk. Most distinguish between small and large animals and animals kept as pets, although this can be difficult, especially with animals like rabbits, which may be raised for meat and pets. They also indicate conditions and if, when, and what kinds of permits are needed.

Zoning for urban livestock generally limits the number and types of animals allowed and addresses noise, odor, and waste management issues, as well as animal welfare concerns like crowding, shelter, and on-site water. It also regulates the types of structures allowed and setbacks from property lines and sidewalks. Beekeeping codes often require permits and "flyway barriers" to stop bees from flying to neighboring properties.

Figure 5.11 Urban chickens.
Source: USDA/FPAC. Photo by Preston Keres.

Seattle, Washington, allows most livestock, poultry, and bees in all zones as an accessory use. Standards are based on lot sizes. Neither roosters nor swine are allowed, except for miniature potbelly pigs. Miniature goats are allowed if they are dehorned and the bucks are neutered. Only one animal is allowed for every 10,000 square feet, and both the animals and their shelters must be kept at least 50 feet from other lots in residential zones. Limits on fowl are extended for community gardens, urban farms, and lots over 10,000 square feet, and farm animals including cows, sheep, and horses are permitted on lots of 20,000 square feet or more. Beekeeping must be registered with the state Department of Agriculture and is limited to four hives of one swarm each on lots less than 10,000 square feet with setback requirements.[40]

Signage

Signs are an important way to promote local farms, especially those that are engaged in agritourism and retail sales. Off-farm directional signs are especially important in rural areas and less traveled roads where farms and ranches are hard to find. Zoning regulations address both permanent and seasonal signs and generally specify their size, placement, and sometimes what materials can be used. They often allow farms to display permanent signs to advertise the business and temporary signs to advertise seasonal products.

Wineries, Breweries, and Distilleries

Wineries, breweries, and distillers can be a hot zoning topic. In cities, they often require a conditional use permit, especially if they are larger than 10,000 square feet. In rural areas, they may be regulated separately or as part of an agritourism ordinance.

Growing and harvesting grapes and grains to produce wine, beer, cider, and spirits generally falls under agricultural production. However, processing, sales, tasting, and events related to creating and marketing alcoholic beverages often are regulated separately. Sometimes some or all of these activities are permitted by right or as an accessory use; other times they require a special permit, especially if the operation hosts large events.

ZONING FOR WINERIES AND BREWERIES IN ALBEMARLE, VIRGINIA

Figure 5.12 Grape vines and photovoltaic panels rely on the sun at Cooper Vineyards in Louisa, Virginia, the first East Coast winery to be awarded Platinum certification by Leadership in Energy and Environmental Design (LEED). The tasting room collects rainwater from the roof and heats and cools the entire building using a geothermal system. Much of the construction materials are from local and recycled material sources.
Source: USDA. Photo by Lance Cheung.

Albemarle County, Virginia's zoning ordinance has a detailed treatment of breweries and wineries. Activities tied to agricultural production are mostly allowed with a complex set of regulatory provisions. Some operational uses are allowed by right, others require a special permit. Helicopter rides and restaurants are prohibited.

Uses permitted by right include:

- Production and harvesting of fruit and other agricultural products;
- Sales, consumption, and tasting of wine and beer;
- Sales of incidental gifts such as corkscrews, bottle openers, glasses, or T-shirts; and
- Private personal gatherings by the owner, provided beverages are not sold or marketed.

So long as they are in accordance with state and federal regulations, the code allows direct and wholesale sales and shipment of wine and beer, storage and warehousing, and uses and sales related to agritourism like hayrides, catering, picnics, and tours, so long as they are limited to 200 or fewer people and do not create a substantial impact on public health, safety, or welfare.

Larger events, such as weddings and receptions, are allowed when clearly associated with agritourism but generally require a special use permit. Considerations include whether the operation has facilities like on-site fermentation, bottling, a tasting room, and a minimum of five acres of agricultural products planted on-site—or on abutting land under the same ownership. Events are subject to zoning clearance if the property is less than 21 acres or if visitors will generate more than 50 vehicle trips/day.

Events with more than 200 people require a special use permit. Considerations include the number of people, the frequency and duration of proposed events, parking availability, a traffic management plan, lighting, and whether there is a stage or structure where music is performed. Finally, the ordinance requires a minimum of 125-foot setbacks in rural districts for structures including tents and portable toilets, as well as off-street parking areas.[41]

Some wineries have gone further to integrate renewable energy systems into their operations. This vineyard (see Figure 5.12) installed photovoltaic panels and collects rainwater from the roof of its tasting room, heating and cooling the entire building using a geothermal system. It even used local and recycled materials for much of its construction.

Discussion Questions

1. Does your state require and/or incentivize comprehensive planning?
2. What kind of zoning and other land use policies are in place in your community?
3. Are they working to protect farmland and/or to support food systems?
4. What other policies or programs are needed?

Notes

1. Freedgood et al., *Farms Under Threat*.
2. Ibid.
3. Oregon Department of Land Conservation and Development, "Oregon Statewide Planning Goals and Guidelines," *Oregon Department of Land Conservation and Development*, July 2019.
4. Oregon Legislative Assembly, "Chapter 215 – County Planning; Zoning; Housing Codes, 2021 Edition," accessed March 15, 2023.
5. "ORS 215.780 – Minimum Lot or Parcel Sizes," Pub. L. No. Oregon Laws, Oregon Administrative Rules, Oregon Revised Statutes, Volume 6, Title 20, Chapter 215, Section 215.780, accessed March 15, 2023.
6. Urban Land Institute, ed., *America in 2015: A ULI Survey of Views on Housing, Transportation, and Community* (Washington, DC: Urban Land Institute, 2015).
7. "Prairie Crossing | A Conservation Community in Grayslake, Illinois," accessed January 29, 2023.
8. Office of Agricultural Services, Montgomery County, MD, "Agricultural Land Preservation," accessed March 15, 2023.
9. Township of Chesterfield, Burlington County, New Jersey, "Township of Chesterfield, NJ: Article XVII Voluntary TDR Program Procedural Requirements," Township of Chesterfield, NJ Code, accessed March 15, 2023.
10. Steven M. Bruder and New Jersey State Agriculture Development Committee, "Data on the TDR Program," December 22, 2022.
11. Alison Davis and Simona Balazs, "The Influence of the Agricultural Cluster on the Fayette County Economy," *Community and Economic Development Initiative of Kentucky College of Agriculture, Food, and Environment University of Kentucky*, May 2017.
12. "Zoning Ordinance of Lexington-Fayette County, Kentucky, Ordinance No. 122–2022, Supp. No. 7, Update 1" (2022).
13. "Zoning Ordinance of Lexington-Fayette County, Kentucky, Ordinance No. 122–2022, Article 1 – General Provisions and Definitions" (2022).
14. Freedgood et al., *Farms Under Threat*.
15. Benjamin Goldstein, Dimitrios Gounaridis, and Joshua P. Newell, "The Carbon Footprint of Household Energy Use in the United States," *Proceedings of the National Academy of Sciences* 117, no. 32 (August 11, 2020): 19122–30.
16. USDA National Agricultural Statistics Service, "Table 1. County Summary Highlights: 2022," 2024.
17. USDA National Agricultural Statistics Service, "2017 Census of Agriculture County Profile: Scott County Iowa," 2017.
18. Scott County Iowa, "Zoning Ordinance for Unincorporated Scott County" (2016).
19. "2007 Scott County Comprehensive Plan," Scott County, Iowa, July 17, 2015; Scott County Iowa, "Zoning Ordinance for Unincorporated Scott County."
20. Chris Mathias, Interview with Chris Mathias, Scott County IA Planning Director, December 12, 2022.
21. "Chapter 336 – Urban Garden District," Ord. No. 208-07 Cleveland Ohio Municipal Code Chapter 336 § (2007).
22. City of Somerville, "City of Somerville Urban Agriculture Zoning Amendment," August 16, 2012.

23. "Chapter 17.20 Agriculture Zone," Eagle Mountain Municipal Code Ordinance O-33-202 § (2023).
24. Anders Van Sandt, Sarah A. Low, and Dawn Thilmany, "Exploring Regional Patterns of Agritourism in the U.S.: What's Driving Clusters of Enterprises?," *Agricultural and Resource Economics Review* 47, no. 3 (December 2018): 592–609.
25. National Agricultural Law Center Research Staff, *Agritourism – National Agricultural Law Center: States' Agritourism Statutes* (Fayetteville, AR: National Agricultural Law Center, February 9, 2021).
26. "Butte County, CA Code of Ordinances: Division 7. – Agricultural Buffers," Chapter 24, Article III, Division 7 §, accessed March 17, 2023.
27. Clark County Washington, "Clark County Code: 40.240.130 Agricultural Buffer Zones in the General Management Area," Title 40: Clark County, Washington, Unified Development Code, October 26, 2022.
28. U.S. Environmental Protection Agency, "Reducing the Impact of Wasted Food by Feeding the Soil and Composting," Overviews and Factsheets, August 12, 2015.
29. U.S. Composting Council, "Model Compost Rules Template (MCRT) Version 2.0," accessed March 15, 2023.
30. Burlington County Agriculture Development and Board, *Burlington County Comprehensive Farmland Preservation Plan (2022 Update)* (Burlington County, NJ: Burlington County Agriculture Development Board, 2022).
31. New York City Department of City Planning, "Zoning Districts & Tools: Rules for Special Areas: FRESH Food Stores -DCP," accessed March 15, 2023.
32. Laura Nixon et al., "Fast-Food Fights: News Coverage of Local Efforts to Improve Food Environments Through Land-Use Regulations, 2000–2013," *American Journal of Public Health* 105, no. 3 (March 2015): 490–96.
33. Julie Samia Mair, Matthew W. Pierce, and Stephen P. Teret, "The City Planner's Guide to the Obesity Epidemic: Zoning and Fast Food," *The Center for Law and the Public's Health at Johns Hopkins & Georgetown Universities*, October 2005.
34. Anna L. Haines, "What Does Zoning Have to Do with Local Food Systems?," *Journal of Agriculture, Food Systems, and Community Development* 8, no. B (October 17, 2018): 175–90.
35. City of Fresno the Development and Resource Management Department, "Fresno Municipal Code Chapter 15: Citywide Development Code" (2016).
36. Hillsborough County City-County Planning Commission, *Food Innovation Districts: A Best Practices Guide to Supporting Locally Produced Agriculture and Food Related Businesses in Hillsborough County* (Tampa, FL: Hillsborough County City-County Planning Commission, n.d.).
37. "Town of Lowville, NY: Anaerobic Digesters" (2013).
38. North Iowa Area Council of Governments, "Cerro Gordo County Ordinance Number 15: Zoning Ordinance of Cerro Gordo County, Iowa: Revised through December 13, 2022," accessed March 15, 2023.
39. Bayfield County Wisconsin, "Bayfield County, Wisconsin Ordinances Title 5, Chapter 6: Large-Scale Concentrated Animal Feeding Operations" (2016).
40. City of Seattle, "Urban Agriculture Ordinance 123378," Ordinance 123378 § (2010).
41. "Albemarle County Code: Chapter 18. Zoning, Section 5: Supplementary Regulations" (2019).

6
PROGRAMS AND POLICIES TO SUSTAIN AGRICULTURE

Chapter Summary

Beyond land use planning and zoning, state and local governments have developed a suite of policies and programs to sustain agriculture and protect agricultural land from urbanization and suburban sprawl. Most states have programs to support agricultural conservation and improve environmental outcomes, including support for local conservation district staff. A handful of states have explored ways to support farm viability and access to land. This chapter describes the types of programs and policies used to protect farmland, support farmers and ranchers, and address farm-related soil, water, and increasingly climate concerns.

***

Beyond land use planning and local regulations like zoning (see Chapter 5), state and local governments plan for and have policies to protect farmland, promote agricultural conservation, and sustain their agricultural sector. However, while the Farm Bill is a comprehensive bill that addresses most aspects of food and agriculture, state policies tend to be piecemeal. Pennsylvania is an exception. It created the first state farm bill in 2019. Fully funded for a fourth time in 2022, it allocated millions of dollars of federal American Rescue Plan funds to agricultural conservation and business development efforts, and provides funding to respond quickly to agricultural disasters, whether animal health, plant health, or foodborne illness.

Every state and many local governments have taken steps to protect farmland from non-farm development. The most common approach is property tax relief. In addition, more than half the states and nearly 100 local governments invest in programs to purchase agricultural conservation easements (or development rights). Some bundle incentives to protect farmland and/or require mitigation if farmland is converted to more developed use. They also support soil and water conservation, composting and reducing food waste, and address water rights. Finally, every state and many local governments provide nuisance protections for generally accepted agricultural practices, and many have taken steps to support agricultural leadership and business planning.

Farmland Protection Policies

In 1981, farmland conversion achieved national recognition with the landmark *National Agricultural Lands Study*. NALS documented the causes and extent of farmland loss and what state and local governments were doing about it. It found development was consuming as much as three million acres of agricultural land annually—nearly a third of which was irreplaceable cropland.[1] While its findings were controversial, few disputed the overall trend: Farmland was rapidly being converted out of agriculture. Since then, state and local governments have stepped up efforts to protect farmland and ranchland for farming and ranching. This section outlines the main farmland protection tools.

Property Tax Relief

Responding early to the pressures of non-farm development, Maryland passed the first law to tax farmland at its current use instead of its speculative value in 1956. Since then, all 50 states and some local governments have enacted agricultural property tax relief. Most have authorized *use value* or *current use assessment* policies to mitigate high land values and property taxes. A few have created circuit breaker or other supplemental tax relief policies.

Use-Value Assessment

Forty-nine states tax agricultural land at its current use value for farming or ranching instead of its potential use for non-farm development. Each program has its own participation requirements and ways to addresses issues like ownership, size, income, and management. Some are stronger than others.

Effective programs have criteria and independent verification to ensure enrolled land remains in agricultural production. In Oregon, landowners are disqualified if their land is removed from an effective-farm-use zone. Other programs assess rollback penalties or require landowners to pay a conversion tax when land is disenrolled or no longer farmed. In Wisconsin, the conversion penalty kicks in when farmland is sold out of agriculture, and a percentage charge is assessed based on how many acres are converted and the market value of the sale. States like Delaware, Maryland, Michigan, and Pennsylvania dedicate a portion of their conversion penalties to permanent farmland protection; in New Hampshire, this is left up to towns. Other programs include provisions to protect and conserve farmland. California, Michigan, and Hawaii require landowners to restrict the use of their land for a specified term, while Massachusetts has a right of first refusal. A few states require conservation adoption or compliance. For example, Vermont can remove farmland and farm buildings if a farm violates state water quality requirements.

Other Tax Relief Policies

Michigan and New York have *circuit breaker tax programs*, which tie property tax relief to income instead of to property values. They provide most relief to farmers with the greatest need. In Michigan, eligible landowners receive a state income tax credit in the amount of property tax paid in excess of 3.5 percent of their total household income. New York offers qualified farmers a state income tax credit for local school taxes. Iowa's Ag Land Credit program partially offsets school taxes by allowing a credit for any general school fund tax in excess of $5.40 per $1,000 of assessed value.

Leasing Development Rights

Some local governments reduce property tax assessments by allowing landowners to lease development rights by placing a *term easement* on their land. This approach complements other property tax relief programs, especially for part-time farmers who do not qualify for agricultural assessment. Southampton, New York, invites landowners to enroll farms of ten acres or more in a local agricultural district and protects them with a ten-year agricultural easement.[2] In exchange, the town grants density and open space set-asides and helps landowners secure economic development funds. Enrolled land also can be used as a sending area for TDR. In Maine, towns can develop a Voluntary Municipal Farm Support Program and grant 20-year agricultural easements in exchange for a tax reimbursement on farmland and farm buildings. The town of Winslow adopted the first VMFSP ordinance in 2016.[3]

Other Exemptions

Some communities provide additional property tax exemptions for farm buildings, forgiving increases in assessed value, which may result from improvements like greenhouses, grain storage, farm labor housing, or value-added processing. Some go so far as to exempt farm equipment and machinery. Others provide sales tax exemptions for specific kinds of farm purchases. New York exempts items from state and local sales and use taxes if they are used "predominantly" for agriculture, including things like building materials, services to install, maintain, or repair farm buildings or structures, motor vehicles, and energy or refrigeration.

Agricultural District/Security Laws

Agricultural districts—aka *agricultural security areas*—package benefits to retain farms and farmland. Created at the state level, some local governments have augmented state programs with their own ordinances. Not to be confused with agricultural zoning (see Chapter 5), agricultural district programs are a comprehensive, incentive-based response to development pressure. Fifteen states and a handful of local governments have active programs with varying acreage requirements and terms of enrollment. Most have links to local planning and are most effective when planning bodies have a role in ensuring their alignment with comprehensive plans. A few include restrictive agreements to protect farmland and/or link to other policies like PACE (see Figure 6.1).

Common benefits include:

- Restrictions on non-farm development, including the use of term easements;
- Use assessment to tax agricultural land at its value for agriculture;
- Protection from public conversion, including actions to limit or prevent eminent domain takings of enrolled land; and
- Protections from nuisance suits and unreasonable local regulations.

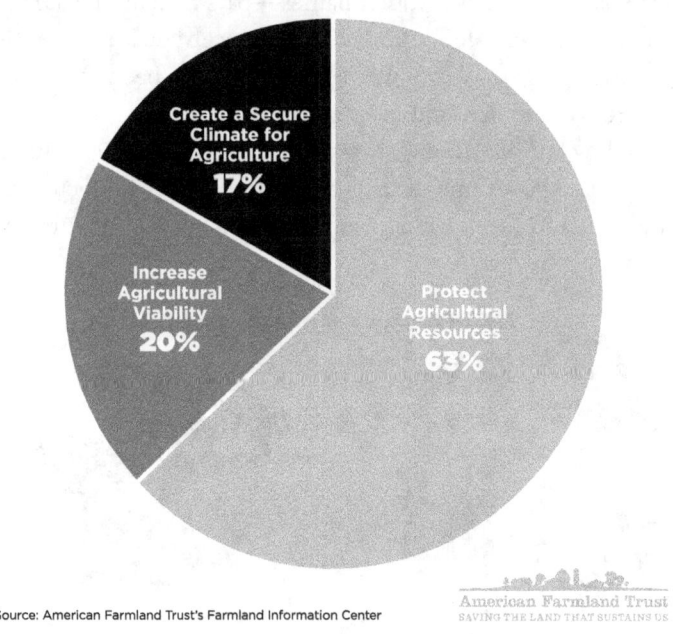

Source: American Farmland Trust's Farmland Information Center

Figure 6.1 Agricultural district programs bundle benefits to retain farms and farmland. Source: American Farmland Trust Farmland Information Center.

California's Williamson Act was the first program, passed in 1965, to protect agricultural resources, preserve open space, and promote efficient urban growth. Local governments enter into rolling, ten-year contracts with landowners in locally designated agricultural preserves. Landowners receive agricultural use assessments in exchange for keeping their land in agriculture or related uses.[4] In 1971, New York followed with Act 25AA, which provides right-to-farm protections for landowners enrolled in state-certified agricultural districts. It also authorizes counties to form agricultural and farmland protection boards to advise local legislative bodies and planning boards on agricultural issues, and to request the commissioner of agriculture to review state agency rules or regulations that affect farming within an agricultural district.[5] Delaware's law, passed in 1991, is linked both to planning and to the state's PACE program (see box).

Purchase of Agricultural Conservation Easements

A Purchase of Agriculture Conservation Easement (PACE)—aka *Purchase of Development Rights* (PDR), among other names—pays agricultural landowners to protect their land from development. First conceived as an incentive-based alternative to zoning, over time PACE programs have been found to support and even strengthen local land use regulations.[6]

Pioneered by Suffolk County, New York, in 1974, by 2022 30 states and nearly 100 local governments had authorized PACE to secure a permanent supply of land for working farms and ranches (see Figure 6.2). Most

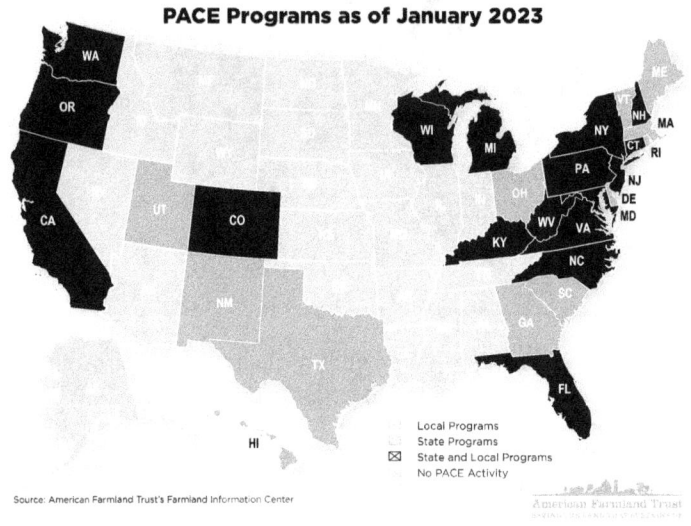

Figure 6.2 Thirty states and nearly 100 local governments have authorized PACE programs to protect farmland.

programs prioritize properties based on soil quality, threat of non-farm development, and future farm viability. Private conservation organizations, like land trusts, also use PACE, often leveraging public funds to support protection. Together, public and private efforts have protected about eight million acres of farmland and ranchland across the U.S.[7]

Traditional easements allow access to or use of someone else's property. They typically are granted to utility companies to provide services like electricity and subdivisions for shared driveways or access to public roads. *Conservation easements* are different, as they allow landowners to limit development to protect natural resources. Voluntary and flexible deed restrictions, they are based on the principle that landowners have a set of property rights that can be donated, transferred, or sold. These include the rights to develop and to limit development on their land (see box). *Agricultural conservation easements* specifically protect land for farming and ranching.

PROPERTY RIGHTS ARE LIKE A BUNDLE OF STICKS

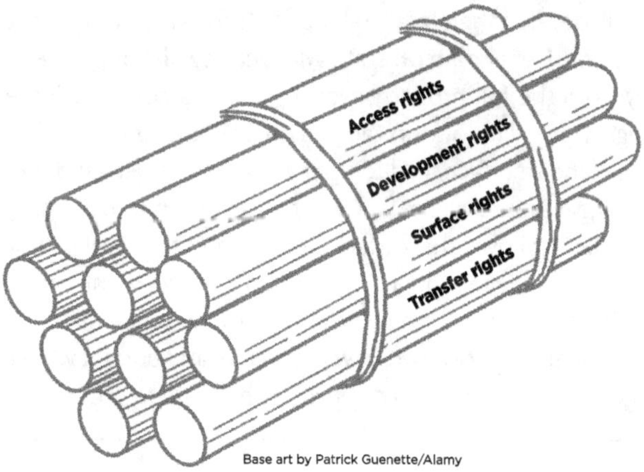

Figure 6.3 Bundle of sticks.

Property rights often are described as a "bundle of sticks" that guide a landowner's legal rights to their property. Each "stick" is associated with a different right, including the rights to use, lease, sell, subdivide, protect, and bequeath property however they like within the limits of local, state, and federal laws. Specifically, these rights include:

- **The right of possession:** When someone holds title to a property, legally it is theirs with all the rights and responsibilities of ownership.
- **The right of control:** Owners control the use of their property within the limits of local zoning, state, and federal laws.
- **The right of access or exclusion:** Owners decide who to allow or prevent from using or entering their property (except for law enforcement if they have a warrant).
- **The right of enjoyment:** Owners can design, develop, use, and live on their property pretty much however they please so long as they respect the rights of neighbors and other community members.
- **The right of disposition:** Owners can sell, rent, gift, or otherwise transfer ownership or use of the property however they want.

In addition, *air rights* allow landowners to occupy vertical space above their property. In rural areas these may be limited to prevent drones or low-flying planes from disturbing poultry and livestock. In urban areas they may be used to regulate the height of buildings or development over highway corridors. *Development rights* give property owners control over whether or how they will develop their land so long as they comply with applicable land use regulations. Depending on state and local legislation, these can be sold separately from the full property, typically using a conservation easement. *Mineral rights* give landowners access to soil and geological formations beneath the land's surface, including metals like copper, gold, iron, and silver, and energy resources like oil, gas, coal, and uranium. *Surface rights* give landowners control over the surface of their land to build, plant crops and graze livestock, produce timber, use water, and so on. In mineral-rich states, landowner's rights are divided between surface and mineral rights. In these cases, landowners may own whatever is on the land but not what is beneath it, although they are allowed to dig wells or build underground storage.

Successful PACE programs enjoy strong public support. They have clear purpose statements, designated funding mechanisms, and make consistent investments in farmland protection. Many are funded through bonds or general appropriations, but funding sources also include lottery proceeds in Colorado; a cigarette tax, municipal landfill fees, and gas-well impact fees in Pennsylvania; and a corporate business tax in New Jersey.[8] NRCS provides matching funds to eligible entities through its Agricultural Land Easements (ALE) program.

Three states and a couple of local programs have augmented their programs by adding mechanisms to keep protected land affordable and in active agricultural use. Massachusetts and Vermont have instated an *Option to Purchase at Agricultural Value* (OPAV). OPAVs require that when protected land is sold, it is sold for a price that reflects its value for farming, not estate or other values influenced by non-farming demand. New York has instated pre-emptive purchase rights. Similar to OPAVs, these are meant to ensure that protected land is sold to qualified farmers. Affirmative easements go a step further to limit land only to agricultural use.

DELAWARE TAKES A HOLISTIC APPROACH TO FARMLAND PROTECTION

Figure 6.4 This area of rural Kent County shows how the Delaware Agricultural Lands Preservation Program has achieved landscape-scale preservation through multiple easement purchases over time. The map shows 134 individual easements purchased since 1996, protecting 43 percent of the land in view on the map. Easements were selected through a voluntary process and prioritized by discounting offered by landowners. The program's success is having a permanent impact on the local landscape and providing land for future generations of farmers.
Source: Map provided by the Delaware Agricultural Lands Preservation Program.

Delaware is the second smallest state, but it ranks first for the value of agricultural sales per acre and second behind California in per-farm sales.[9] To save farmland for food and other products, the state legislature created the Agricultural Lands (Aglands) Preservation Program in 1991, combining farmland protection and land use planning. Professional planning staff advise and support the AgLands Preservation Foundation, which uses PACE and agricultural districts to retain farmland and support farm enterprises at a landscape level. Together, these programs have permanently protected more than 25 percent of the state's farmland and enrolled 400 farms in districts, totaling more than 40,000 acres.

Landowners participate through a two-phase process. In Phase I, they sign a ten-year agreement to enroll in an Agricultural Preservation District and commit their land to farming. Farming is defined broadly. It includes traditional crop, livestock, and poultry production, but also agritourism, equine, forestry, and non-commercial hunting, trapping, and fishing. Enrolled landowners receive nuisance protections and notification requirements for adjacent landowners, and their unimproved land is exempt from county, school, and real estate transfer taxes. Farms must be zoned for agriculture, meet income requirements, and achieve a minimum farm viability score of 170 using an automated Land Evaluation and Site Assessment (LESA) model. LESA is NRCS's numeric rating system used to evaluate a parcel's long-term viability based on soil type, surrounding land use, zoning, and other attributes.

After a year, landowners may apply to Phase II to sell an agricultural conservation easement through the state's Aglands Preservation program. If sold, the district agreement is replaced by a permanent easement. Easements are selected through a voluntary process and prioritized by discounting offered by landowners. Funding permitting, the Aglands Program selects one round of farms to protect each year.[10]

Farmland Mitigation

A handful of state and local governments require developers to offset the impacts of developing farmland by protecting an equal or greater amount of comparable land. Mitigation can be achieved by requiring contributions to a PACE program, purchasing land in fee, and/or dedicating funds to a qualified conservation entity.

In Massachusetts and Vermont, if there is no feasible alternative to developing high-quality farmland, developers must contribute funds to their state PACE programs. In California, more than a dozen local governments require farmland mitigation. Yolo County requires developers whose projects convert prime farmland in specified zones to protect at least three acres of farmland for every acre converted. Projects that convert lesser-quality soils must protect a minimum of two acres, and projects that convert a combination must mitigate at a blended ratio. Affordable housing developments are exempt, as well as public uses, such as parks and schools, that do not generate revenue. California also has a program that provides grants to buy agricultural easements in counties surrounding railway corridors.[11]

Promote Conservation

Agricultural conservation includes land management practices that maintain or improve land and water resources for the future to advance sustainable environmental and economic use. Beyond Farm Bill programs administered by FSA and NRCS, state and local governments have programs to support soil health, water quality and quantity, on-farm conservation programs, composting, and reducing food waste. Increasingly, they also are confronting a changing climate.

In 2014, California took agricultural conservation a step further by combining it with farmland protection and land use planning to combat climate change through its Sustainable Agricultural Lands Conservation Program (see box). It also created a Tribal Government Challenge planning grant program to invest in tribal efforts to support climate-related feasibility studies, planning activities, and auditing to advance tribally tailored solutions to reduce GHG emissions and improve access to clean energy while advancing climate adaptation and resiliency on tribal lands.[12]

CALIFORNIA'S SUSTAINABLE AGRICULTURAL LANDS CONSERVATION PROGRAM UNITES PLANNING AND PROTECTION

Figure 6.5 Chino, California, agricultural landscape.
Source: USDA.

California Senate Bill (SB) 862 established the Affordable Housing and Sustainable Communities (AHSC) Program "to reduce greenhouse gas emissions through projects that implement land use, housing, transportation, and agricultural land preservation practices to support infill and compact development."[13] Led by the cabinet-level Strategic Growth Council, AHSC is funded by proceeds from the state's cap-and-trade emissions reduction program. Its farmland component, Sustainable Agricultural Lands Conservation (SALC)[14] supports land use planning and permanent farmland protection to promote smart growth, reduce vehicle miles traveled, and increase community food security.

SALC is administered by California's Department of Conservation and the Natural Resources Agency. They fund communities to protect farmland and encourage compact, transit-oriented development with three types of awards:

1. Planning grants to support local and regional land use policies, economic development strategies, and plans to protect critical agricultural land. Plans also may support environmental co-benefits of protecting farmland and ranchland.

2. Agricultural Conservation Acquisition grants to buy conservation easements on or fee titles to agricultural land.
3. Capacity and Project Development grants to expand organizational capacity to develop agricultural conservation acquisition projects.

SALC has had significant impacts. By the end of 2022, it had protected 196,000 acres of farmland and ranchland, provided planning grants to 30 communities, and capacity grants to 20 more. And it estimates that so far, its GHG reductions (over 30 years) will be over 20.6 million metric tons of carbon dioxide equivalent.[15]

Soil Health Policies

The Dust Bowl stimulated the first generation of soil conservation. As understanding of the benefits of soil health have evolved, the focus is shifting from solving erosion to enhancing ecological performance and climate resilience. As of 2022, 20 states had passed soil health resolutions or legislation.[16] Many initiatives provide producers and landowners with technical and financial assistance to implement and/or demonstrate conservation practices to

Figure 6.6 Healthy soil has a rich, deep color. Use of a diverse blend of crops, grasses, and cover crops creates a protective blanket that feeds and nurtures the soil.
Source: USDA-NRCS. Photo by Catherine Ulitsky.

improve soil health, sequester carbon, and reduce GHG emissions. Some fund capacity building for conservation districts, local governments, and other service providers to help farmers and landowners advance soil health.

California has taken the most comprehensive approach. Beyond SALC, its Healthy Soils Initiative is a collaboration of state agencies and departments that work together to promote soil health on farmland and ranchland. It offers financial assistance to producers to implement soil health practices and funds demonstration projects to showcase results. A handful of states have created soil health tools, as well as soil health testing and monitoring programs. For example, Illinois's Saving Tomorrow's Agricultural Resources (STAR) program offers a voluntary framework to assign points for best management practices. States like Colorado have adopted STAR and created state soil health testing and monitoring programs (see box).

TWO STAR PROGRAMS IN COLORADO PROMOTE SOIL HEALTH

Figure 6.7 Improving soil health on America's rangeland can yield significant water quality improvements, sequester more carbon, and improve forage for livestock.
Source: USDA.

In 2021, the Colorado state legislature established a voluntary Agricultural Soil Health Program. It authorizes the Colorado Department of Agriculture to establish a soil health testing program and monitoring system, and to provide grants and loans to farmers and ranchers to advance soil health practices. It also creates a diverse 11-member soil health advisory committee comprising conservation district board members, farmers and ranchers, a tribal member, water users, and three ex officio members.

The program is supported by state stimulus funds and grants from the state Department of Public Health and the Environment, the Colorado Water Conservation Board, National Fish and Wildlife Foundation, and NRCS. It has two main parts:

- Saving Tomorrow's Agricultural Resources (STAR) is a simple framework originally developed by the Champaign County Soil and Water Conservation District in Illinois. It helps farmers and ranchers assess their current systems, identify ways to improve soil health, measure progress, and share success. It assigns points for management activities. Scores are converted to a 1–5 STAR Rating, with 5 STARs demonstrating commitment to a suite of practices that improve soil health, water quality, and water availability.
- STAR+ provides financial and technical assistance to producers to adopt new practices on one field over three years and to consider expanding them across their entire operations. It also provides capacity support, equipment grants, training, and other incentives to conservation districts and eligible entities to provide technical assistance to producers and landowners.[17]

In September 2022, USDA awarded Colorado a $25 million Climate Smart Commodities grant to expand the reach of the STAR+ program and help other interested western states adopt the STAR evaluation system.[18]

On Farm Conservation Programs

American Farmland Trust's Farmland Information Center found 37 states have programs to encourage conservation adoption. Many are administered by state departments of agriculture or conservation commissions, often in partnership with local conservation districts. Most offer technical assistance, including direct consultations and technical expertise on developing conservation plans and applying conservation practices. Some provide training and support to help producers and landowners comply with environmental regulations (see box). Many provide public funds to farmers, ranchers, and agricultural landowners to implement practices to improve soil and water quality. Funding sources range from general state funding, real estate transfer taxes, cap-and-trade proceeds, and grants from public and private sources.

While these programs increasingly address soil health practices to mitigate climate change, given strict Clean Water Act requirements, on-farm conservation programs generally address livestock management. For example, the North Carolina Agricultural Cost Share Program is a typical voluntary, incentive-based program to address nonpoint source pollution. Local conservation district staff work with agricultural landowners and renters to develop and approve conservation plans, identify and design practices to ensure longevity, and acquire approval of a cost share agreement to install best management practices.[19]

LOUISIANA MASTER FARMERS LEARN TO IMPROVE WATER QUALITY

The Louisiana Master Farmer Program is a multi-agency effort to help farmers address environmental concerns while enhancing production and resource management skills. After EPA placed several of the state's water bodies on its impaired waters list, the state legislature enacted the program to reduce pollution entering its rivers and streams, empowering the commissioner of agriculture and forestry to certify individuals as master farmers.

Farmers must complete a training program and implement a comprehensive soil and water conservation plan that meets the standards and specifications of the Louisiana Department of Agriculture and Forestry, affected soil and water conservation districts, and NRCS.[20] LSU AgCenter, the state's land-grant university, developed the voluntary program in collaboration with key partners. Available to all Louisiana farmers on a watershed-by-watershed basis, it teaches environmental stewardship, conservation-based production, and sustainability through classroom instruction and a hands-on field day and workshop.

Building Capacity to Support On-Farm Conservation

Along with direct support to landowners and producers, states build the capacity of technical service providers like conservation districts, local governments, and other entities. Capacity-building programs fund staff, program delivery, and in some cases offer specialized training. For example, New Mexico's Healthy Soil Act supports agricultural systems and other land management to improve the health, yield, and profitability of the soils of the state. The state's Department of Agriculture awards grants to conservation districts and other qualified entities, including Extension and land-grant universities, Acequias, Pueblos, tribes, and nations to implement on-the-ground projects to achieve soil health principles.[21] Kansas's Water Resources Cost-Share Program allocates funds to counties to address locally determined water resource concerns. Funding can be used for a range of approaches, including terraces and waterways to address soil erosion. It also can be used to create new springs and other water sources to keep cattle away from streams and rivers. New York's Climate Resilient Farming Grants program awards competitive grants to conservation districts in three project categories: agricultural waste storage cover and flare for methane reduction, on-farm water management, and soil health systems.[22]

Clean Water Indiana (CWI) supports technical and financial assistance to both landowners and conservation groups. Under the direction of the State Soil Conservation Board, it is funded by a portion of the state's cigarette tax. It provides grants to conservation districts to address priorities like nonpoint sources of water pollution. Districts use these funds for locally driven education and capacity-building programs, including cost share programs, staffing for technical assistance, equipment, field days and outreach programs. Grants also support regional specialists to help counties address invasive species concerns. Many projects target farmers who currently are not served by other conservation programs.[23]

Water Quality Policies

The Clean Water Act requires states to set water quality standards to ensure waterways are safe for drinking, recreation, and other designated uses. Waterways that fail to meet the standards are listed as impaired and the maximum amount of a pollutant allowed in a waterbody—or Total Maximum Daily Load (TMDL)—must be developed to bring them into compliance.

EPA has authority to regulate point pollution from sources like CAFOs, but states have authority over nonpoint pollution, such as nutrient runoff from farmland. They use a series of approaches to minimize the impacts of nonpoint pollution on water quality. These include limiting application of agricultural nutrients, nutrient management plans, tracking and reporting, and financial incentives like water quality trading or income tax credits.

Most states regulate applications of manure, fertilizers, and other agricultural nutrients and limit how, where, and when producers may apply nutrients to land. Some restrictions are based on weather conditions and time of year, others on setbacks and buffers, and still others on application methods.[24]

Snow and frozen ground increase the risks of manure runoff and water contamination. So Iowa passed a law that prohibits CAFOs with more than 500 animals from applying liquid manure in certain winter conditions. These include on ground covered with 1 inch of snow or 0.5 inches of ice from December 21 to April 1, and on frozen ground from February 1 to April 1, unless soil is frozen to a depth of 2 inches or less.[25] Texas law forbids applying nutrients when the ground is frozen or saturated, including heavy rainfall events. It requires CAFOs to base application on soil analyses and apply them uniformly to suitable land at appropriate rates and times according to specific crop needs. It also requires buffers to protect water supply wells used for irrigation, and between application areas and sinkholes or water—with exceptions for alternative conservation practices or field-specific conditions.[26]

Pennsylvania's Resource Enhancement and Protection (REAP) program takes an incentives-based approach. It provides state income tax credits to farmers, landowners, and businesses to offset implementation costs of practices to reduce nitrogen, phosphorus, and sediment pollution. REAP offers farmers tax credits of 50–75 percent of a project's eligible out-of-pocket costs; 90 percent for farmers who operate in a watershed with an EPA-mandated TMDL. In 2019, REAP was amended to allow farmers to earn up to $250,000 in tax credits to buy equipment or cover crop seed, or to conduct soil health testing.[27]

VERMONT RAPs TO ADDRESS NONPOINT SOURCE POLLUTION

Figure 6.8 This Vermont dairy conserved sensitive riparian areas on their farm by establishing forested buffer zones and installing high-tensile fence, stream crossings, and other water handling equipment.
Source: USDA.

Vermont's Clean Water Act—Act 64—requires the state's department of agriculture to set Required Agricultural Practices Rules (RAPs) with farm management standards to address nonpoint source pollution and nutrient losses and set construction and siting requirements for farm structures in floodplains, floodways, river corridors, and flood hazard areas. Act 64 governs a wide range of practices from livestock management and crop production to storage and handling of agricultural wastes and fuel. It also regulates things like ditching and subsurface drainage, construction and maintenance of farm infrastructure, and the stabilization of farm fields adjacent to banks of surface water.[28]

RAPs apply to farmers who meet a low bar of minimum threshold criteria. Criteria include producing an average of $2,000 or more of gross annual agricultural income, and/or farming 4.0 contiguous acres or more, and/or managing specified numbers of livestock (e.g. four horses

> or 250 broilers). They require both vegetated buffer and riparian buffer zones. Buffers must maintain 10–25 feet of perennial vegetation between croplands and adjacent surface waters and ditches. Tillage and mechanical application of agricultural wastes are prohibited in the zones.
>
> Operations that meet the criteria and manage manure, other agricultural wastes, or fertilizers must implement a nutrient management plan.[29] For other operations, nutrient application rates must be accounted for and based on current university recommendations and standard farming practices. Fields that receive mechanical application of manure, fertilizers, and other nutrients must be soil sampled at least once every five years and operators must maintain records showing compliance.[30]

Nutrient Management Plans

Almost every state has laws and regulations requiring producers to develop written nutrient management plans to address the impacts of point and nonpoint source pollution. Nutrient management plans address the type, amount, timing, and application of manures, fertilizers, and other soil amendments. They typically include results from soil tests to determine crop needs, an inventory of nutrient sources, and best management practices for applying manure and minimizing run off. Most of these laws focus on animal feeding operations to respond to the Clean Water Act's National Pollutant Discharge Elimination System (NPDES). Some require plans for all agricultural nutrients, and some have established independent programs. Virginia's Pollution Abatement Permit Program regulates animal waste and other discharge to surface waters that fall beneath the state's NPDES permit thresholds. Animal feeding operations—including poultry—must implement an approved nutrient management plan, including a site map showing where applications occur, a site evaluation and assessment of soil types, a description of activities, application schedules, land requirements, soil and waste monitoring, and annual reporting.[31]

Water Quality Trading

The federal Clean Water Act caps how much pollution can flow into waterways. This has led several states to develop trading programs to buy and sell pollution reduction credits. Water quality trading is a market-based approach that allows pollution dischargers to buy nutrient reductions from other

PROTECTING WATER QUALITY IN THE CHESAPEAKE BAY

Figure 6.9 Chesapeake Bay Watershed.
Source: Edwin Remsberg/USDA-SARE.

Maryland has several agricultural conservation programs to protect and restore the Chesapeake Bay and its tributaries. Its Agricultural Nutrient Management Program is the most comprehensive, requiring certified preparers to develop nutrient management plans for most farms in the state. Plans include all farming practices related to nutrient use, identify what and how nutrients will be used and disposed of, and recommend ways to manage fertilizers and other nutrient inputs. Operators file annual reports explaining what they did and certifying they will follow the plan the next year. Plans are updated every three years, with some exceptions, and Maryland's Department of Agriculture conducts on-farm audits to verify that farmers are following their plans.[33]

Maryland's Agricultural Certainty Program gives farmers a ten-year exemption from new environmental laws and regulations in exchange for installing best management practices that meet TMDLs ahead of schedule. A certified verifier inspects farm applicants every three years to determine compliance with local, state, and federal environmental requirements. The Chesapeake Bay Watershed water quality trading program provides a public market for nitrogen, phosphorus, and sediment

> reductions. Farms that meet nutrient reduction requirements can generate credits from practices such as cover crops, reduced nutrient application, and riparian buffers. Credits can be purchased by point sources and other interested buyers through a nutrient trading market.

sources. Cities, farms, and facilities that discharge wastewater to waterways can meet regulatory requirements by buying equivalent or larger pollution reductions from another source or by protecting or restoring floodplains, riparian areas, and wetlands.

The Ohio River Basin Water Quality Trading Project is the largest water quality trading program. Led by the Electric Power Research Institute and a diverse group of collaborators, it funds farm conservation practices in Ohio, Indiana, and Kentucky. It is supported by watershed modeling, on-the-ground project verification, and a credit registration that includes state agency verification. Credits are given for verified nitrogen or phosphorous reduction, which counties can sell. They also can be purchased as a bundle of benefits that include nutrient reductions and other ecosystem benefits, such as GHG reductions and pollinator or endangered species protections.[32]

Water Rights

Agriculture depends on water and often requires irrigation. Farms with some form of irrigation account for more than half the value of U.S. crop sales.[34] Irrigation accounted for 42 percent of U.S. freshwater withdrawals in 2015.[35]

Producers source water from surface water—rivers, lakes, streams, and ponds—and from groundwater stored in aquifers. Water quantity—or allocation—is mostly governed by the states. Policies address uses such as withdrawal from aquifers and water bodies, and transfers between watersheds. Typically, they regulate water based on its source, addressing surface and groundwater separately, and creating a complex tapestry of rules, regulations, and interstate competition.

Different regions use different allocation systems to determine private water rights. The eastern U.S. generally relies on riparian rights, the west on prior appropriation, while a handful of states like California and Oklahoma use a hybrid approach.

Riparian rights are broad but limited to landowners adjacent to the water body they plan to use. Landowners may make "reasonable use" of the water as long it does not interfere with rights of other riparian landowners. Irrigation and other agricultural uses are considered reasonable uses under most laws. Most riparian states use a permitting system to plan and optimize water use. A state agency issues the permits and controls the who, how much, and when of water usage. In many states, agriculture is exempt from permit requirements.

Prior appropriation is a "first-in-time, first-in-right" seniority system. Priority is given to the longest-standing appropriator who gets their water first. Water users can take their full appropriative right until the water is gone, even during times of drought. While these laws vary from state to state, they all are based on diverting water from a natural course and include general requirements about applying water to beneficial uses, including agriculture. As with riparian rights, most states use a permitting system.[36]

Groundwater allocation policies often differ from surface water and rely on multiple legal doctrines. They generally differentiate between on-tract use where water is sourced where it is located, and off-tract use where it is not. Many western states use prior appropriation for groundwater as well as for surface waters. In most New England and a few other states, landowners are allowed to use as much ground water as they want without regard to neighboring users. Some states have tempered this with correlative rights to distribute water more equitably. Other states require water to be put to reasonable use on overlying land but do not allow it to be transported. Some states combine approaches. Finally, some states allow and even encourage rainwater collection by offering tax credits or exemptions for the purchase of harvesting equipment, while others, especially western states, regulate rainwater collection.

In regions where irrigation is prevalent, local irrigation and groundwater organizations often manage on-farm water withdrawals and help manage irrigation pumping from local groundwater resources. For most, delivering water to farms is their primary function, but they have dual functions as well. Some perform drainage and groundwater management functions, others address wildlife management and flood retention (see Figure 6.10).

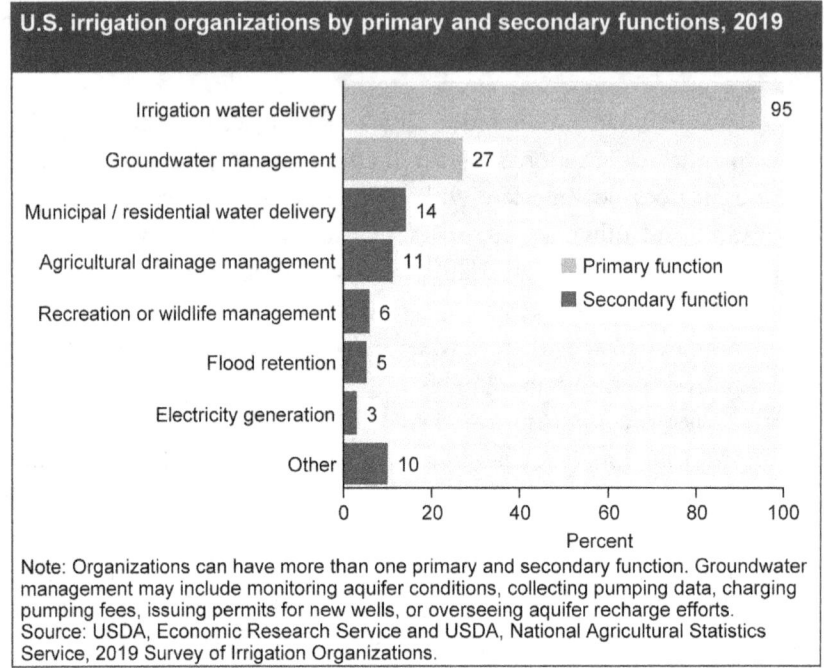

Figure 6.10 Irrigation organizations have various functions, but the main one is delivering water.

Composting and Food Waste

USDA estimates that between 30 and 40 percent of food is wasted annually in the U.S.[37] Most of it ends up in landfills where it produces methane, a noxious greenhouse gas.[38] States have begun to address this with composting policies. Compost is created by combining organic wastes from food, manure, yard trimmings, and other organic matter into containers, piles, or rows. When managed in the right ratios, it creates nutrient-rich humus, which can be used in place of fertilizers to promote higher crop yields and remediate damaged soils. It also helps sequester carbon.

Vermont was the first state to ban food waste in landfills. Composting was found to be the main disposal method used to comply with the law.[39] California's Senate Bill 1383 was enacted to lower methane emissions from landfills and to rescue food for people to eat. It set targets to reduce organic waste disposal 75 percent and rescue 20 percent of currently disposed surplus food by 2025. Local governments are required to collect food and other organic waste from residents and businesses and to produce clean streams of organic feedstock that can be recycled into compost, renewable natural gas, electricity, and paper. The law includes manure, biosolids, digestate, and sludges, plus green material, landscape and pruning waste, organic textiles and carpets, lumber, wood, and paper products.[40]

Other states have composting rules that address basic operating and location requirements to protect water quality and prevent nuisance complaints. These tend to be geared toward large-scale operations and address management issues such as turning piles or windrows and measuring temperature. Some states exempt farms from these restrictions. In Wisconsin, farms composting crop residue, manure, and on-site farm animal carcasses are exempt from most state composting requirements.[41]

Increasingly, local governments are adopting composting programs. Most operate like municipal trash and recycling services with curbside pickup. Some support backyard and community programs and educate residents on how to compost on their own. Others require composting for supermarkets and other businesses that produce quantities of organic waste. Portland, Oregon, requires businesses that assemble, cook, process, serve, or sell food to compost food scraps. This includes stores, restaurants, hotels, and cafeterias, food and beverage manufacturers, hospitals, corporate and college campuses, and starting in 2024, elementary and secondary schools engaged in food preparation. Exceptions are made for businesses that donate food to a charitable organization, such as a food bank or pantry. They also are made for scraps used to feed animals in facilities regulated by the state's Department of Agriculture and collected for rendering or biofuel production.[42]

Support Agricultural Viability

Agricultural viability builds on the concept of "farm viability," or an individual farm's ability to sustain long-term production and succession. It goes a step further to consider a jurisdiction's ability to maintain economically sustainable farm enterprises, support wealth creation, and retain land in agricultural production for future generations. State and local governments have developed a suite of policies to enhance agricultural viability and sustain their agricultural sectors. Some support agricultural leadership and farm business planning. Others protect farmers from conflicts with new neighbors. Still others offer bonds to help beginning farmers get started.

Right-to-Farm Laws and Ordinances

Since 1963, when Kansas enacted the first law to protect feedlots from litigation, all 50 states have passed Right-to-Farm laws to protect farmers from unreasonable regulations and nuisance complaints. Nuisance protections are especially important when new residents move into agricultural areas with unrealistic expectations of country living and then object to the routine noises, dust, smells, and practices associated with commercial farm production.

Right to Farm provisions also may be included in zoning enabling laws and agricultural district programs. Some local governments have enacted their own ordinances to strengthen and clarify language in state law and to educate residents about agricultural activities. For example, Sonoma County, California's states:

> No agricultural operation conducted or maintained on agricultural land in a manner consistent with proper and accepted customs and standards, as established and followed by similar agricultural operations in the county, shall be or become a nuisance . . . if it was not a nuisance when it began, provided that such operation complies with the requirements of all applicable federal, state, and county statutes, ordinances, rules, regulations, approvals and permits. The provisions of this section shall not apply where a nuisance results from the negligent or improper management or operation of an agricultural operation.[43]

New Unit Notifications

New unit notification ordinances warn prospective buyers and new residents of the types of farming activities they should expect. They direct realtors or landowners to notify potential buyers of properties next to or near farms of right to farm or other policies that might affect them. Some require the ordinance be placed in public areas and/or periodically mailed to residents to illustrate local support for agriculture.

> **LOCAL AG DISTRICT'S ORDINANCE REQUIRES PUBLIC NOTIFICATION**
>
> To protect farming areas for local food and fiber production, Northampton County, North Carolina, requires certified properties to change their titles to state that the property is located within one-half aerial mile of an agricultural district. Developers and others applying for building permits must sign a statement indicating that they have reviewed the agriculture district map and that they understand the types of activities that take place in these districts. Agricultural district maps must be posted along with a public notice in key county offices, including the Register of Deeds and the Planning Department, and include the following language:
>
>> Northampton County has established Voluntary Agricultural Districts to protect and preserve agricultural lands and activities. These districts have been developed and mapped by the county to inform all purchasers of real property that certain agricultural and forestry activities, including but not limited to pesticide spraying, manure spreading, machinery and truck operations, livestock operations, sawing, and other common farming activities may occur in this district any time during the day or night.[44]

Agricultural Advisory Boards and Commissions

Many communities have created formal or informal advisory boards to serve as the voice of agriculture in local affairs. These may be created at the local or regional level and used to serve as ambassadors to the public, identify and resolve issues of concern to farmers, and ensure local policies value and support agriculture.

Some states have created voluntary *Agricultural Commissions* to provide a voice and engage farmers in developing local policies and programs to support agriculture, protect farmland, and encourage agricultural businesses. Typically, these are advisory boards to give farmers a voice in community decision-making. However, in California they play a formal and regulatory role. There, commissioners are charged with protecting the environment and public health and safety, while advisory boards play a role similar to commissions in other parts of the country. In addition, several California counties have created a *Farmbudsman* position to help farmers navigate the regulatory process.

The size and composition of these commissions and boards vary by location and may include members of other local boards with related interests, such as a planning commission, zoning board, conservation district, or economic development commission. In Connecticut, most agricultural commissions have five to seven members who are either farmers or involved in farm-related businesses. In King County, Washington, the Agricultural Advisory Committee comprises nine people appointed by the county Board of Supervisors from various agricultural industries, the county Farm Bureau, and interests including water, small farms, processing, agricultural equipment, agricultural chemicals, and petroleum.

Aggie Bonds

Sixteen states offer aggie bond programs to help beginning farmers acquire land, pay for improvements, and buy things like equipment and livestock.[45] Aggie bonds are established through a federal–state partnership and generally run by a state department of agriculture or financing authority. They allow private lenders to receive federal and/or state tax-exempt interest on loans made to beginning farmers. This makes it possible for local lenders to offer lower interest rates on loans. The interest the beginner pays on the loan is exempt from federal—and in some cases, state—income taxes.

Agricultural Economic Development Authorities

Some state and local governments have created economic development authorities to make capital, loans, and other financial support available to producers. Most programs are run out of state departments of agriculture, but some are independent authorities. For example, the Maryland Agricultural and Resource-Based Industry Development Corporation (MARBIDCO) is a quasi-public corporation. It was broadly authorized in 2004 by Maryland's governor and general assembly to develop agricultural industries and markets; support commercialization of agricultural processes and technology; assist with rural land preservation efforts; and alleviate the shortage of nontraditional capital and credit available at affordable interest rates for investment in agricultural and resource-based businesses. At the local level, Polk County, North Carolina, created an Agricultural Economic Development office to encourage and expand the county's agricultural enterprises. Staff assist in training and scaling up agricultural businesses, identify trends and support services, create marketing opportunities, and connect new farmers to mentors. Building on the county's comprehensive plan, it also works to protect farmland to support a viable agricultural economy.

Cooperative Extension

The Smith-Lever Act established the Cooperative Extension System in 1914 to educate rural Americans about advances in farming practices and technology. It is a cooperative partnership between federal, state, and local governments to share scientific research from land-grant universities and—among other things—support farmers and ranchers with production and business technical assistance.

Extension is funded through annual Congressional appropriations administered through USDA's National Institute for Food and Agriculture. Federal dollars must be matched with state and local appropriations as well as grants, contracts, and fees. Federal funds cover anywhere between 10 and

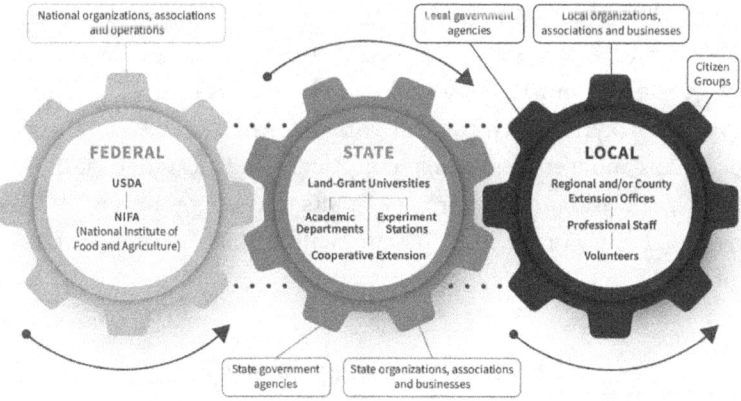

Figure 6.11 Cooperative Extension empowers farmers, ranchers, and communities to meet the challenges they face, adapt to changing technology, improve nutrition and food safety, prepare for and respond to emergencies, and protect our environment.

50 percent of its services, states between 20 and 50 percent, and counties up to 20 percent.

Farm Viability Programs

Four states have programs that provide teams of experts to help producers with business planning, and in some cases farm transfer planning and funding for capital improvements. Enacted in 1996, the Massachusetts Farm Viability Enhancement Program was the first. It offers one-on-one business planning, technical assistance, and implementation grants of up to $150,000 to improve farm viability and to preserve and support the stewardship of agricultural resources. Funding can be used for capital projects such as building or renovating barns, farmstands, and other retail structures; increasing food storage and processing capacity; wash-pack facilities; and so on. It also may be used for farm equipment and/or improving farm infrastructure, such as wells or fencing. Farmers who receive grant funding must sign a term easement to keep their land in agricultural use for a ten- or 15-year contract.

Market Support

State and local governments have many ways to promote agriculture and expand market opportunities. Most have procurement policies to encourage schools and sometimes other institutions to give preference for purchasing products from local farms. Many spearhead campaigns like Jersey Fresh or Kentucky Proud, which support state-wide marketing efforts, and produce maps and websites to help consumers find local food, agritourism, and U-Pick operations. Some provide grants to cover producers' marketing expenses or create, invest in, and staff facilities like farmers' markets and food hubs.

"Buy Local" Branding Campaigns

Many states and regions have developed "buy local" and other branding campaigns to encourage consumers to buy more food and farm products from the state or region where they live. These campaigns add value to local products and bolster local economies. They use a variety of branding messages and communication tools, from billboards to social media to events, and often publish annual directories to connect consumers with local farms. Some states provide grants to support these campaigns.

Farmers' Markets

Farmers' markets are organized, often temporary places where farms can sell products directly to consumers. In 2022, nearly 10,000 farmers' markets operated in communities across the U.S.,[46] supporting a growing demand for local food while improving the viability of small and mid-sized farms and addressing food insecurity by bringing fresh produce to underserved communities. Many state and local governments make grants to support farmers' markets, invest in infrastructure to start or maintain markets, or license and handle permitting of vendors.

Food Hubs

Food hubs are middle-market infrastructure to manage aggregation, distribution, and marketing of local and regional farm products. Generally, private businesses and organizations as well as state and local governments have supported their development to strengthen local agriculture and address food insecurity, often using USDA local food marketing funds. Hawaii went further and in 2022 passed legislation requiring its Department of Agriculture to establish a five-year food hub pilot program. It provides grants to establish or expand a food hub or provide technical assistance. Local governments also support food hubs with grants, loans, and tax rebates, streamline permitting processes, and work with agriculture to support certification standards.

Geographic Preference Procurement Policies

Farm-to-school programs catalyzed many state and local governments to create geographic preference procurement policies to support farm viability and increase access to local food in public institutions like schools, hospitals, and prisons. Today, most states have farm-to-school policies with programs to advance core elements, including local procurement, school gardens, and food and agriculture education.[47]

Figure 6.12 Girls promoting a farm-to-school program in New York.
Source: Josh Baldo.

Some places have gone further to require public agencies to purchase a specified amount of food from local farms. Others create pilot programs or take a hybrid policy/program approach. Cleveland, Ohio, gives 2 percent bid discounts to businesses that are sustainable, locally based, or purchase 20 percent of their food locally. These can be combined for a maximum discount of 4 percent.[48] New York City directs city agencies to give preference to products grown, produced, or harvested in the state. Guidelines apply to solicitations valued at more than $100,000 for food or food-related services, and for social services through which more than $100,000 of food would be purchased annually in fulfillment of the contract.[49]

Discussion Questions

1. What kind of support does agriculture need in your community?
2. What kinds of policies or programs might contribute to that support?
3. Who from the agriculture community might you engage in your planning process?

Notes

1. U.S. Department of Agriculture and President's Council on Environmental Quality, *National Agricultural Lands Study* (Washington, DC: National Agricultural Lands Study, January 1, 1981).
2. "Southampton, NY: Lease of Development Rights Enabling Ordinance – (Southampton, NY, Code § 330–248G (2023)," Farmland Information Center.
3. Maine Department of Agriculture, Conservation & Forestry, "Voluntary Municipal Farm Support Program: Farmland Protection Program: M.R.S. 7 Ch2-C," accessed March 23, 2023.
4. For more information about the Williamson Act, visit: "Williamson Act Program Overview."
5. For more information about New York's agricultural districts law, Act 25AA, visit: "Chapter 69: Agriculture and Markets, Article 25-AA: Agricultural Districts," Consolidated Laws of New York § (2017).
6. Thomas L. Daniels, "Assessing the Performance of Farmland Preservation in America's Farmland Preservation Heartland: A Policy Review," *Society & Natural Resources* 33, no. 6 (June 2, 2020): 758–68.
7. Farmland Information Center, *Local and State Purchase of Agricultural Conservation Easement Fact Sheets and 2022 Land Trust Survey* (American Farmland Trust, 2023).
8. Julia Freedgood et al., "Farms Under Threat."
9. "Delaware: A Small State That Is Big in Agriculture."
10. "Aglands Preservation Program," Delaware Department of Agriculture – State of Delaware, and Julia Freedgood et al., "Farms Under Threat."
11. "Sec. 8–2.404. Agricultural Conservation and Mitigation Program.," Title 8, Chapter 2, Section 8–2.404 Yolo County Code of Ordinances §.
12. State of California Governor's Office of Planning and Research and California Strategic Growth Council, "Grant Funding Opportunity: Tribal Government Challenge Planning Grant Program," March 13, 2020.
13. Committee on Budget and Fiscal Review, "Senate Bill 862," Pub. L. No. 862 (2014).
14. Committee on Budget and Fiscal Review, "Senate Bill 862"; California Strategic Growth Council, "Affordable Housing and Sustainable Communities."
15. Shanna Atherton-Bauer, "California Climate Investment Programs and Farmland Protection," December 23, 2022; California Climate and Action Network, "Sustainable Agricultural Lands Conservation Program (SALCP)."
16. Nerds for Earth, "State Healthy Soil Policy Map," *Nerds for Earth*, August 27, 2019.
17. Colorado Department of Agriculture, "Soil Health."
18. John A. Miller, email March 17, 2023.
19. North Carolina Department of Agriculture & Consumer Services, "Soil & Water Conservation Division – Cost Share Programs: Agricultural Cost Share Program (ACSP)."
20. "RS 3:304 Master Farmer Certification," Louisiana Laws – Louisiana State Legislature § (2008).
21. New Mexico Department of Agriculture, "Healthy Soil Program."
22. New York State Department of Agriculture and Markets, "Climate Resilient Farming."
23. Clean Water Indiana, "Clean Water Indiana: 2023 Competitive Grants," October 25, 2022.

24. Peggy Kirk Hall and Ellen Essman, *State Legal Approaches to Reducing Water Quality Impacts from the Use of Agricultural Nutrients on Farmland* (Fayetteville, AR: National Agricultural Law Center, May 2019).
25. Iowa Department of Natural Resources, "Winter Manure Application: Guide to Frozen and Snow-Covered Ground Rules," May 2021.
26. Texas Commission on Environmental Quality, "Subchapter B: Concentrated Animal Feeding Operations §§321.31–321.47," Chapter 321 – Control of Certain Activities by Rule § (2014).
27. Pennsylvania General Assembly, "Act of Jul. 25, 2007, P.L. 373, No. 55 Cl. 72 – Tax Reform Code of 1971 – Omnibus Amendments" (2007); Pennsylvania State Conservation Commission, "PA REAP FY 2019 Annual Report."
28. Vermont General Assembly, "Title 6: Agriculture, Chapter 215: Agricultural Water Quality, Subchapter 2: Water Quality; Required Agricultural Practices and Best Management Practice Including §§ 4810, 4810a, and 4811."
29. USDA Natural Resources Conservation Service, "Conservation Practice Standard Nutrient Management (Code 590)," May 2019.
30. *Vermont Required Agricultural Practices Rule for the Agricultural Nonpoint Source Pollution Control Program (Act 64 of the Vermont General Assembly, 2015 Session)* (Montpelier, VT: Vermont Agency of Agriculture, Food & Markets, November 23, 2018).
31. "Virginia Administrative Code – Title 9. Environment – Agency 25. State Water Control Board – Chapter 32. Virginia Pollution Abatement (VPA) Permit Regulation."
32. Jessica Fox, "Trading Up: The Ohio River Basin Water Quality Trading Project Is the World's Largest Water Quality Credit Program," *Water Technology*, November 2, 2019.
33. Maryland Department of Agriculture, "Agricultural Nutrient Management Program," Maryland.gov Enterprise Agency Template.
34. USDA Economic Research Service, "Irrigation & Water Use," May 6, 2022.
35. Cheryl A. Dieter et al., "Estimated Use of Water in the United States in 2015," *USGS Numbered Series, Estimated Use of Water in the United States in 2015*, vol. 1441, Circular (Reston, VA: U.S. Geological Survey, 2018).
36. For a more detailed overview of water law, visit the National Ag Law Center: National Agricultural Law Center, "Water Law Overview."
37. U.S. Department of Agriculture, "Food Waste FAQs."
38. U.S. Environmental Protection Agency, "Reducing the Impact of Wasted Food by Feeding the Soil and Composting," *Overviews and Factsheets*, August 12, 2015.
39. Emily Belarmino et al., "Impact of Vermont's Food Waste Ban on Residents and Food Businesses," *College of Agriculture and Life Sciences Faculty Publications*, January 30, 2023.
40. For more information about California's law, visit: "California's Short-Lived Climate Pollutant Reduction Strategy," CalRecycle.
41. Wisconsin Department of Natural Resources, "Composting Rules and Regulations in Wisconsin."
42. City of Portland, OR, "Business Food Scraps Requirement," April 3, 2022.
43. "Sonoma County, California. CHAPTER 30 – AGRICULTURE. Article II. – Right to Farm," Ord. No. 5203 § 5, 1999 §.
44. "Northampton County, NC Voluntary Agricultural District Ordinance."

45. Council of Development Finance Agencies, "CDFA – CDFA Spotlight: Aggie Bonds," CDFA.
46. National Farmers Market Directory, "Find a Local Farmers Market Near You."
47. National Farm to School Network, "State Farm to School Policy Handbook 2002–2020" (National Farm to School Network, Vermont Law School's Center for Agriculture and Food Systems, July 2021).
48. Growing Food Connections, "Cleveland, OH Local Purchasing, Ordinance No. 1660-A-09 |," Growing Food Connections Policy Database.
49. New York City Mayor's Office of Contract Services, "New York State Food Purchasing Guidelines, 2012 (Revised April 7, 2015)."

7

PROGRAMS AND POLICIES TO SUPPORT COMMUNITY FOOD SECURITY

Chapter Summary

Most funding for nutrition assistance comes from federal Farm Bill programs but is administered by states. Many states have augmented federal programs with nutrition incentives to support healthy food choices for people in need. State, local, and tribal governments also support healthy retail policies, nutrition education and promotion, and emergency food programs that give away surplus food through food banks, pantries, soup kitchens, and other feeding sites. This chapter offers examples of various policies and programs that address food insecurity and improve access to healthy foods.

***

As discussed in Chapter 1, food insecurity is most widespread in rural areas and large inner cities. Most funding for nutrition assistance comes from federal Farm Bill programs, primarily SNAP with highest participation in rural areas.

SNAP is administered by states. States determine eligibility, issue monthly benefits, and otherwise direct implementation. Some have expanded participation, while others have passed laws to limit it by adding eligibility rules and disqualifying people for things like failing to pass a drug test.[1] States also cover healthy meals under Medicaid.

Local governments generally act as on-the-ground representatives to explain how SNAP works and help eligible recipients apply. Ten states have delegated SNAP administration to counties so they assume a larger role.[2]

Many states and several tribes augment federal programs like FMNP, SFMNP, and GusNIP to supplement the purchasing power of low-income households and increase consumption of fresh fruits and vegetables (see Chapter 4). USDA Food and Nutrition Service reported more than 3,100 farmers markets accepted SNAP[3] in 2022 and several state as well as local governments require farmers' markets to accept EBT.

In 2020, an average SNAP beneficiary received about $121 per month—about $4.00 per day[4]—often too little to support a well-balanced diet. States have responded in various ways:

- Offering farmers' markets no-cost EBT equipment to increase SNAP redemption;
- Extending benefits to additional populations, such as income-eligible students;
- Staggering benefit distribution to attract retailers to neighborhoods where SNAP is widely used;
- Contributing state dollars to increase recipients' purchasing power; and
- Partnering with SNAP-Ed to educate recipients about healthy food choices.

Beyond their roles with federal nutrition programs, state, local, and tribal governments support other policies and programs to increase access to fresh fruits, vegetables, and other nutritious foods. Increasingly they are considering ways to ensure food availability in the event of supply chain disruptions—especially those that occur during pandemics, weather disasters, and other emergencies. While planning departments often are involved, these efforts also are supported by Cooperative Extension, health departments, civic organizations, and *Food Policy Councils* that convene government agencies and diverse food system stakeholders to promote public health and the social and economic benefits of local and regional food systems.

Most policies aim to increase food access to priority populations who experience barriers to securing ingredients for a healthy diet. They support farmers' markets and other outlets for supplying fresh produce, including mobile markets, produce carts, and government-supported food stalls. They enact geographic preference policies to purchase local food for institutions (see Chapter 6), fund food pantries and school meal programs, and provide land for urban agriculture and community gardens. Several have either eliminated food taxes or begun to tax junk food and soda. *Healthy retail policies* address the needs of residents who lack grocery stores and other retail outlets close to where they live. They include several kinds of programs, including healthy food financing. *Nutrition Incentive Programs* augment federal benefits under WIC and SNAP and promote consumption of fruits and vegetables. *Nutrition education and promotion* programs help residents select and prepare foods to improve health outcomes. Finally, *emergency food programs* give surplus food away through food banks, pantries, soup kitchens, and other feeding sites.[5]

Food Policy Councils

The Johns Hopkins Center for a Livable Future's Food Policy Networks project maintains a comprehensive online directory of food policy councils across North America. Also called alliances and networks, food policy councils build capacity and convene stakeholders to address food-related issues and needs, including food access, food security, and food procurement. Some go further to address economic development, food production and processing, and even land use planning. Typically, they operate at the municipal or county level, some at the regional (municipal-county, multi-county, or multi-state), and the rest at state or tribal levels.[6]

Food policy councils sometimes are created by legislation, but most are quasi-independent, with public agency staff playing a role. Some or all members may be appointed by a government agency, involve agency personnel, or the council may receive financial support from a state or local government. Thus, some councils are embedded in or sanctioned by a governmental body, but many operate independently.

Most often, they play educational and networking roles. They collect and share knowledge and data with public officials, inform the public, and lead community engagement efforts. This was especially evident during Covid-19 when 82 percent reported facilitating connections across sectors to match food and farm resources, tapping their knowledge and networks to share information widely on food systems' needs.[7] Some work to start new programs, improve coordination between existing programs, and/or advocate for public policies.

Councils generally focus on locally important issues. Sometimes their proposals are included in comprehensive plans. In Sarasota County, Florida, a food policy council collaborated with the county planning department to facilitate seven public meetings. Using a World Café activity, participants generated 39 food policy recommendations. Access to healthy food and support for sustainable agriculture were later incorporated into the county's comp plan recommendations and adopted by County Commissioners.[8] The Arizona Food System Network successfully advocated a statewide "Double Up Food Bucks" nutrition incentives program. During Covid, the council worked to gain additional emergency funding for Double Up Arizona as well as the local farm–to–food bank program. It also released Arizona's first statewide Food Action Plan to help guide decision-making and inform policy opportunities.[9]

Healthy Retail Programs

ERS estimates that in 2021, 34 million Americans lived in food-insecure households without ready access to fresh, healthy, and affordable food.[10] Most of them were low income and lived in low-resource communities that lacked full-service grocery stores or supermarkets. This has costly public health implications. According to the CDC, leading causes of preventable diseases include poor nutrition, tobacco, and excessive alcohol use. Communities as well as individuals are harmed when stores do not carry fresh fruits, vegetables, and other healthy foods but make cigarettes, highly processed "junk" food, and alcohol widely available.

Health disparities are especially harsh for Black/African American communities, especially in urban census tracts where they have the fewest supermarkets of any racial or ethnic population.[11] Various social and economic factors discourage groceries from locating in low-resource areas, including a history of redlining, lack of viable sites, costly development procedures, and negative perceptions about community residents and their spending power.[12]

Healthy retail policies address these barriers to improve access to nutritious food in underserved communities. They use a range of tactics and strategies to attract supermarkets, create new grocery stores, and help existing stores stock, market, and sell fresh produce, low-fat milk, whole grains, and other healthful foods. Among other things, they use nutrition incentives to increase buying power and tools like taxes, licensing, and zoning to balance the ratio of healthful to junk food in these communities.

Healthy Retail Programs improve public health outcomes and strengthen local economies. Stores that accept SNAP and WIC bring federal dollars to communities. Stocking a variety of healthful food options gives storeowners a competitive edge. Further, supermarkets and groceries attract other businesses, creating jobs, increasing tax revenues, and stabilizing home values.[13]

Healthy Corner Store Initiatives

Convenience and corner stores are intended for quick stops to pick up food, drinks, household supplies, and personal care products. However, in many

Figure 7.1 Nyssa Entrekin, a registered dietician with The Food Trust, delivers nutrition education at Olivares Food Market in Philadelphia.
Source: Dave Tavani for the Food Trust.

underserved communities they are the primary food outlet: 15 percent of SNAP purchases are made at these stores.[14] Mostly independently owned, they have limited physical space, available inventory, and refrigeration, so they rarely supply the full range of foods necessary to support a healthy diet.

Given the need for refrigeration, it can be prohibitively expensive for small stores to stock fresh produce and other perishable items. *Healthy Corner Store initiatives* address this barrier by helping retailers pay for infrastructure improvements to increase the availability of healthier food choices in underserved communities. They use several approaches to encourage bodegas, corner stores, and other small convenience stores to stock fresh fruits and vegetables, low-fat dairy products, lean proteins, and other healthy food options. Incentives include funding for equipment, purchasing subsidies, training and technical assistance, and marketing and consumer incentives to drive customers to revitalized stores.

San Francisco, California's Healthy Food Retailer Incentives Program is administered by the city's Economic and Workforce Development Department. It provides stores with targeted technical assistance, consultation, business development tools, and access to programs to strengthen their operations.[15] Healthy Food Retailers are broadly defined to include grocery stores, corner and convenience stores, farmer's market, and other retailers whose business mostly comprises sales of food and other groceries. It supports retailers who devote at least 35 percent of their selling area to fresh produce, whole grains, lean proteins, and low-fat dairy products, but no more than 20 percent to tobacco and alcohol products, and satisfies the minimum wage requirements for employees.[16]

Some places go further than food to promote active living. Through its Mass in Motion Healthy Market Program, the Massachusetts Department of Public Health works with local governments and community-based organizations to help people eat better and move more. It offers grants to cities and towns to create policies, systems, and environments to promote wellness and healthy living. Its Municipal Wellness and Leadership Grant Program provides grants and technical assistance to cities and towns to engage in community-based obesity prevention efforts. These include increasing access to local fresh foods through farmers' markets and working with food retailers to offer healthy, affordable food and beverage options, creating safer neighborhoods with bike lanes and walkable paths, and expanding parks, playgrounds, and other places to be active.[17]

HEALTHY CORNER STORE INITIATIVES

The FOOD TRUST

Figure 7.2 The Food Trust logo.

The Food Trust is a Philadelphia-based nonprofit and nationally recognized leader in healthy retail initiatives. It piloted a Healthy Corner Store Initiative (HCSI) in 2004. Subsequently partnering with the Philadelphia Department of Public Health, the program expanded rapidly, first throughout the city, then to Camden, New Jersey, and by 2014 it had grown to 11 more cities supported by city, state, federal, and private funding sources. Today, nearly 100 stores participate. The program was so successful it has been replicated across the country and USDA published a Healthy Corner Stores Guide as part of its ongoing efforts to help SNAP recipients make healthy food choices.[18]

Nearly a third of Philadelphia's residents participate in SNAP.[19] Many suffer from food insecurity and rely on corner stores to buy their food. HCSI's goal is to increase food access and encourage consumption of fresh fruits, vegetables, and other nutritious choices in low-income communities affected by health disparities. It achieves this by increasing awareness and availability of healthy foods in corner stores through a multifaceted approach:

- Expanding store capacity to sell and market healthy items;
- Offering training and technical assistance to store owners;
- Promoting health messages and providing nutrition education;
- Linking corner store owners to community partners, local farmers, and fresh food suppliers; and
- Offering free blood pressure checks and referrals by health care providers to customers in select corner stores enrolled in the Trust's companion "Heart Smarts" program. The stores also receive in-store nutrition education lessons that include cooking demonstrations and free taste tests.

A peer-reviewed evaluation found HCSI increased store capacity to market and sell healthy food. It significantly increased the availability of fruits, vegetables, and low-fat milk, especially in stores that received upgraded infrastructure, such as refrigeration, shelving, and kiosks to make fresh produce the focal point of the store. It also had economic development impacts. Stores reported increased customer traffic and weekly profits. It was estimated that the HCSI's overall economic impact over 30 months was more than $1 million of earnings, $140,000 of added tax revenue, 38 jobs, and increased property values in neighborhoods with participating stores.[20]

Healthy Food Financing Initiatives

Healthy food financing initiatives (HFFI) are public–private partnerships used to attract retail food establishments and invest in food-access infrastructure in underserved communities. They provide one-time grants and loans to develop or renovate outlets such as bodegas, corner stores, grocery stores, farmers markets, and mobile markets. Usually seeded with state funding, they are administered with a Community Development Financial Institution (CDFI) and a Food Access Organization (FAO).

The Food Trust spearheaded the first HFFI in Pennsylvania in 2004 with support from a state representative and a CDFI called the Reinvestment Fund. It showed that food retailers could overcome the high costs of land, infrastructure, and workforce development to operate profitable healthful-food retail businesses in underserved communities. Launched with a three-year, $30 million state appropriation to the Pennsylvania Department of Community and Economic Development, it provided grants and loans to food retailers to build or expand healthy food markets in lower-income, underserved urban and rural communities. The original program operated until 2010, supporting close to 90 projects across the state, creating or retaining nearly 5,000 jobs and improving access to healthful food for an estimated half a million residents. Pennsylvania reinvested in the program in 2018 and it continues today.[21]

Pennsylvania's success has been replicated in other states and cities and led to the creation of two federal programs. One is funded through the CDFI Fund at the Department of the Treasury. The other was authorized in 2014 and 2018 Farm Bills and is administered by the Reinvestment Fund on behalf of USDA Rural Development. It provides capacity building and financial resources to eligible healthy food retail projects and food supply chain enterprises to overcome high costs and barriers to entry. It expanded significantly with $22.6 million from the American Rescue Plan Act of 2021 and other relief legislation as part of a new USDA framework to transform the food system.

HFFI programs operate at the local level, as well. The New Orleans Fresh Food Retailer Initiative (FFRI) is a partnership between the city, the HOPE Enterprise Corporation, and the Food Trust to stimulate supermarket and grocery store development in underserved communities of Orleans Parish. It provides forgivable and interest-bearing loans for capital, real estate, and related expenses to encourage supermarkets and other retail outlets to expand healthy food offerings in low- or moderate-income neighborhoods. FFRI is partially funded by a $7 million grant to the city and state from HUD's Disaster Community Development Block Grant program and matched one-to-one by HOPE, a credit union. Funds can be used to create, renovate, or expand retail outlets to sell fresh produce while providing underserved neighborhoods with opportunities for employment and revitalization.[22]

Healthy Retail Licensing

Some communities take a regulatory approach to food retailer licensing. Commonly used to curtail tobacco sales, the approach also has been used to establish a baseline of healthy products that food retailers are required to carry. For example, Minneapolis, Minnesota, created an ordinance that requires licensed grocers to carry specified food items to improve the city's health outcomes (see box). Licensing also may include incentives to reward stores that surpass minimum stocking requirements.[23]

A NOVEL STAPLE FOODS ORDINANCE IN MINNEAPOLIS

Recognizing the correlation between food insecurity, low consumption of fruits and vegetables, and poor health outcomes, the city of Minneapolis adopted a Staple Foods Ordinance to help ensure that all city residents had access to healthy foods no matter where they shopped. It required licensed grocers to carry an assortment of healthful foods and beverages to better align with the CDC's dietary guidelines.[24] Licensed grocers included supermarkets and groceries, corner stores, and co-ops, as well as most gas stations, dollar stores, and pharmacies, with some exemptions.

Approved in 2008, the City Council amended the ordinance in 2014 to set clearer and more complete standards. It amended it again in 2018 to align with cultural dietary preferences and to require stores to stock a specified minimum of specified dairy, fruits, vegetables, whole grains, proteins, and other healthy food options on a continuous basis.[25] Based on stakeholder feedback, the changes reduced the number of required food categories as well as some quantities while expanding varieties and package sizes in others.

City staff provide training and resources to help stores understand the regulation and supply the required foods. They offer merchandising and marketing trainings, in-store promotional supplies, and consultations with retail and marketing experts. They also help connect retailers with procurement options and offer low-interest loans for coolers and freezers. Beyond providing incentives, city health inspectors monitor compliance as part of routine inspections. If a store is out of compliance, inspectors write a violation order and instruct store owners to fix the problem. In instances of ongoing non-compliance, they may issue a formal citation and monetary fine, after which a business's license may be revoked.[26]

Mobile Food Vending Allowances

Mobile vending involves selling food out of buses, carts, trailers, trucks, roadside kiosks, and other portable vehicles.[27] Common in urban and rural communities alike, and commonly regulated, mobile vendors supply all kinds of foods—some of it healthy, some of it not. However, with the right allowances and regulations, mobile markets are a convenient way to increase access to healthy foods while both responding to and expanding a community's food culture.

Mobile vending allowances can supply underserved neighborhoods long before a supermarket can overcome cost and regulatory barriers. It can be targeted to areas with large numbers of SNAP and WIC recipients to take advantage of nutrition incentives. Mobile vending can bring food to areas where people commonly congregate. Kansas City, Missouri, has instituted guidelines to increase access to healthier food and beverages in their parks. They have two categories of vendors: "Healthier" and "Healthiest," which must comply with a set of nutritional standards. Healthier vendors receive a 50 percent discount in the cost of a mobile vending permit if half their food meets nutritional guidelines. Healthiest vendors receive a "roaming" permit for the cost of a standard permit if 75 percent of the food they sell meets those guidelines. Vendors are routinely monitored by KC Parks to assure full compliance.[28]

A ROLLING FARMSTAND BRINGS HEALTHY FOOD TO UNDERSERVED NEIGHBORHOODS IN WASHINGTON, D.C.

The Arcadia Center for Sustainable Food and Agriculture launched a "rolling farmstand" in an old school bus in 2012. A decade later, its mobile market has sold about $1.5 million worth of affordable, high-quality, local food in underserved neighborhoods in Washington, D.C., providing over $1 million of wholesale revenue to local farms. Every week the bus brings fruits, vegetables, eggs, dairy, meat, and bread to neighborhoods with low car ownership located at least a mile from a supermarket or fully stocked grocery store. To expand food access further, the program doubles the face value of SNAP, FMNP, and Senior FMNP.[29]

Nutrition Incentive Programs

Thirty states have established fruit and vegetable incentive and produce prescription programs to increase the ability of low-income consumers to buy fresh fruits and vegetables at farmers' markets, other direct marketing outlets, and increasingly grocery stores. A trailblazing Massachusetts program laid the foundation for nutrition incentive policies. Spearheaded by August (Gus) Schumacher, then commissioner of agriculture, the Massachusetts Farmers' Market Coupon program offered coupons to WIC recipients to purchase fresh produce at farmers markets. Piloted in 1986 as a partnership between the state Department of Agriculture and the Federation of Massachusetts Farmers' Markets, it expanded in 1987 to include senior citizens at nutritional risk.[30] In 1992, Congress authorized the federal Farmers Market Nutrition Program (FMNP) to provide fresh, locally grown fruits and vegetables to WIC participants and to expand sales at farmers' markets.

Schumacher went on to work for the World Bank, served as a USDA undersecretary, and co-founded the Wholesome Wave Foundation. Throughout his life, he championed programs to serve the needs of local farmers and low-income populations with limited access to healthy food. Today, the Gus Schumacher Nutrition Incentive Program (GusNIP) and other nutrition incentive programs address food insecurity, improve health outcomes, and strengthen local agriculture in nearly every state as well as several tribes and U.S. territories. Building on this approach, the Fair Food Network launched a pilot "Double Up Bucks" program to match SNAP dollars in five Detroit farmers' markets

Figure 7.3 Gus Schumacher.
Source: TedXMahhatten.

in 2009. As of 2021, Double Up was available at more than 1,325 grocers and farmers' markets across 30 states, including 826 sites where farmers sold produce directly to consumers.[31] SNAP recipients are automatically eligible.

Honoring Schumacher's legacy, Massachusetts created a *Healthy Incentives Program* (HIP) to provide a dollar-for-dollar reimbursement when SNAP users buy fresh local food directly from Massachusetts farmers to support. A state-funded nutrition benefit, HIP is operated by the Massachusetts Department of Transitional Assistance. Between 2017 and 2022, over 175,000 SNAP households purchased more than $42 million of fresh produce from local farmers. Of these, 44 percent of the families included seniors, 30 percent included children, and another 30 percent included a person with a disability.[32]

DOUBLE UP BUCKS TAKES OFF IN OKLAHOMA

In 2020, Hunger Free Oklahoma received a $500,000 GusNIP grant to expand Double Up Oklahoma (DUO) in farmers' markets, and to pilot a grocery store program in rural areas. It matched the GusNIP grant with foundation support and added more than $400,000 from the state's Tobacco Settlement Endowment Trust.

The program was so successful that in 2022, DUO spent nearly $3 million helping more than 14,000 households purchase fruits and vegetables each month. It finalized a new, industry-leading Learning Management System for SNAP enrollment assistance training. And it received another GusNIP grant—this time for $14.2 million—to expand DUO to 44 counties over the next four years. TSET continues to support the program, and the state legislature stepped in with a $1.1 million annual appropriation.[33]

Produce prescription programs—or Veggie Rx—encourage healthy eating to prevent and manage chronic disease. Health care professionals provide "prescriptions" along with EBT or coupons to buy fresh fruits and vegetables. Often, they are facilitated by local health departments or community-based organizations; some states, like Washington, have formally established programs. In 2019, Washington's state legislature established definitions and eligibility and appropriated funds to support a Veggie Rx subprogram under its broader fruit and vegetable incentives program.

The Marion County Public Health Department in Indiana has a county-wide Veggie Rx program to prevent and manage chronic diseases while improving nutrition security. It issues electronic benefits called Healthy Savings, which participants can spend at local Kroger, CVS, and Walgreen stores. During Covid it incorporated virtual nutrition and chronic disease management education.[34] Navaho Nation has a similar Fruit and Vegetable Prescription Program (FVRx) and partners with health care providers and local retailers to promote healthy eating. Eligible families receive a monthly prescription in the form of a voucher to buy fruits and vegetables at stores on Navajo Nation. These stores are encouraged to offer a variety of fruits and vegetables, including locally grown produce from Navajo farmers. It also has a Healthy Navajo Stores Initiative to increase the amounts of fruits, vegetables, and traditional Diné foods available in small stores on Navajo Nation.[35]

Nutrition Education and Promotion

The federal government has a long history of helping citizens and institutions make informed decisions about food both eaten and served. USDA

Figure 7.4 Elementary student showing classmates how carrots grow.
Source: USDA.

and the U.S. office of Health and Human Services (HHS) review and update national *Dietary Guidelines for Americans* every five years. These guidelines inform many federal food provision and nutrition education initiatives, including WIC, the National School Lunch Program, and the Expanded Food and Nutrition Education Program (EFNEP).

State, local, and tribal governments can reinforce or supplement these guidelines by providing their own mechanisms for encouraging and promoting healthy dietary choices. County Extension offices deliver local nutrition education programs. For example, North Carolina's Flagship Program educates people to increase food security and improve health outcomes and works to foster economic development through local and regional food and farming systems. Among other programs, it trains Extension agents about cultural awareness in cooking, how to open food donation stations and connect with a SNAP-Ed so agents can use resources like garden kits, a $500 garden start-up fund, and the Color Me Healthy curriculum to provide garden and local food training.[36] Outside of Extension programming, the most common ways local governments support healthy eating behaviors is by establishing local nutrition guidelines, food marketing policies, or school wellness programs.

Indian Tribal Organizations (ITOs) or state government agencies administer USDA's Food Distribution Program on Indian Reservations, which provides USDA Foods to income-eligible households living on reservations and to Native American households living in approved areas near reservations or in Oklahoma. Many Indigenous households participate as an alternative to SNAP because they lack easy access to SNAP offices or authorized food stores. ITOs and other administering agencies determine applicant eligibility, store and distribute food, and provide nutrition education. USDA provides the administering agencies with funds for program administrative costs.

Nutrition Guidelines

State and local governments can promote healthy eating and increase access to healthful foods by developing nutrition guidelines for public agencies, partner organizations, schools, childcare centers, and health care providers. Generally, these are based on the USDA/HHS Dietary Guidelines for Americans, but they also use other guidelines, like the Harvard Healthy Eating Plate[37] or even their own.

Figure 7.5 The Choctaw Nation offers cooking classes to demonstrate healthy meals that may be prepared with food available through the USDA FNS Food Distribution Program on Indian Reservations.
Source: USDA. Photo by Bob Nichols.

Cleveland, Ohio's Healthy Eating Committee partners worked with the American Heart Association to develop a set of Healthy Cleveland Nutrition Guidelines, which the city adopted in 2014. Spearheaded by the city's Department of Public Health, community partners included the Cleveland Foodbank, Children's Hunger Alliance, Hunger Network of Greater Cleveland, City of Cleveland, Healthy Cleveland, Cuyahoga County, Cleveland Clinic, Ohio State University Extension, and The Cleveland Cuyahoga County Food Policy Coalition. Their goal was to align with USDA Dietary Guidelines and consumer recommendations to improve healthy eating and the health and well-being of greater Cleveland residents by creating clear nutritional guidelines. The guidelines are meant to inform which food will be purchased, donated, prepared, and served by individuals and families as well as by local government, agencies, and organizations. Organizations that receive public food program funding must follow the guidelines, and Healthy Cleveland partner organizations also use them as a framework when they promote and provide food to their clients.[38]

Regulating Food Marketing and Sales

State and local governments can regulate food marketing and sales to promote healthy eating and influence dietary choices. Many places now require calorie information on menus. Some tax sodas and junk food. Still others ban or restrict trans fats or marketing junk food to children. Generally used to discourage negative behaviors, some also fund healthy eating incentives.

New York City was the first local government to enact a calorie labeling law in 2008. The state of California quickly followed in 2009. Subsequently, a provision of the federal Affordable Care Act requires restaurant chains with 20 or more U.S. locations to post the calorie content of prepared foods on menus alongside the item's price. A handful of localities levy taxes on sugar-sweetened beverages—aka "soda taxes"—which generally apply to beverages other than sodas like iced teas and sports drinks. In 2014, the Navaho Nation passed the Healthy Diné Nation Act, which levies a 2 percent tax on junk foods of minimum to no nutritional value sold on Navajo Nation. Proceeds are used to support community wellness programs.[39] Seattle, Washington, started taxing sugar-sweetened beverage products in 2018. Instead of charging a sales tax directly on consumers, its Sweetened Beverage Tax makes distributors pay a tax (of 1.75 cents per ounce) on sugar-sweetened beverage products they distribute within the city.[40]

School Wellness Policies

The Child Nutrition and WIC Reauthorization Act of 2004 has a Local School Wellness Policy requirement. This was strengthened by the subsequent Healthy, Hunger-Free Kids Act of 2010. Schools that participate

Figure 7.6 School lunch staff preparing food.
Source: Josh Baldo.

in the National School Lunch Program and/or School Breakfast Program must develop a local school wellness policy to promote student health and address childhood obesity. Responsibility for developing the wellness policy is placed at the local level but must meet basic standards. Policies must be developed by a collaborative community process, contain nutrition guidelines, be monitored and evaluated regularly, and contain goals for nutrition education, nutrition promotion, and physical activity.

BURLINGTON VERMONT PROMOTES HEALTH MEALS FOR BETTER LEARNING

School districts have the authority to ensure that children in their jurisdictions have access to food and knowledge of nutrition. However, not all have the same capacity or give it the same priority. Burlington, Vermont, takes a comprehensive approach. The city school district partnered with three key community organizations to form the Burlington School Food Project. With a motto of "healthy meals for better learning," it provides local food to district schools, educational opportunities in school gardens, cooking contests for middle and high school students, and cooking classes for both students and food service staff—all with a focus on fresh, local ingredients. For students, these educational opportunities are often integrated with the core curriculum, such as math and geography. And its leading educational program, Fork in the Road, is a student-run food truck that serves free school breakfast, lunch, and summer meals to all students in the district.[41]

Local governments also play a role in ensuring that schools have strong school wellness policies. For example, the Washington, D.C., Council unanimously passed the Healthy Schools Act in 2010. This landmark legislation is meant to improve the health, wellness, and nutrition of District children by helping schools, students, and families eat healthy, stay active, learn healthy habits, care for the environment, and create healthy school communities. Meals must meet the USDA nutrition guidelines, and all public schools are required to serve free breakfast to all students and free lunch to all students who qualify for federal assistance. The program solicits input from students, faculty, and parents to design nutritious meals, and posts information about food served in the school office and on the school website. Schools that meet these requirements receive financial assistance to offset costs, and schools that source food from local farms are eligible to receive an extra five cents extra for each lunch served that contains at least one locally grown, unprocessed meal component.[42]

Emergency Food Systems

In the midst of Covid-19, Feeding America estimated that 53 million people turned to food banks and community programs to put food on their tables. The largest hunger-relief organization in the U.S., Feeding America is a network of 200 food banks and 60,000 food pantries and meal programs that provide food and services to people in need.[43] These organizations provide emergency food free of charge to low-income populations through public and private hunger relief programs. Outlets include food banks, soup kitchens and pantries, homeless shelters, and Meals on Wheels. Although emergency food providers usually are private, local governments can help

Figure 7.7 In the wake of Hurricane Harvey, a pallet of disaster food boxes leave the Houston Food Bank commodity warehouse for delivery to food pantries to help those in need in Houston, TX, on September 22, 2017.
Source: USDA. Photo Lance Cheung.

to ensure that these services are available, well-coordinated, and provide wholesome foods.

Funding for Emergency Food

USDA's Emergency Food Assistance Program (TEFAP) provides food at no cost to low-income people through state distributing agencies. Administered by FNS, TEFAP provides food to states based on the number of residents and unemployed people who have incomes below the state poverty level. In turn, states provide food to local agencies, usually food banks, which distribute it to local organizations, like soup kitchens and food pantries. TEFAP also provides administrative funds to support storage and distribution of USDA Foods. These funds must—at least in part—be passed down to local agencies.

State and local governments can fund their own or build on TEFAP to fund or operate distribution programs. Many support publicly operated services like Meals on Wheels, which feeds senior citizens through home distribution and senior meal sites. Some fund private nonprofit agencies to provide emergency feeding services. For example, in 2022 Seattle awarded $2.8 million to support 22 community-led projects through its Department of Neighborhoods' Food Equity Fund. These projects created new food pantries, community meal programs, intergenerational nutrition workshops, and more. The Food Equity Fund was developed in 2021 to increase investments in community work led by those most impacted by food and health inequities. It is supported by the city's Sweetened Beverage Tax and City Council funds.[44]

During Covid-19, local governments became more involved in emergency food provision. Cities including Atlanta, Georgia; Columbus, Ohio; and Madison, Wisconsin, used general or special funds to support distribution efforts. Others used funding from the Coronavirus Aid, Relief, and Economic Security (CARES) Act. Dekalb County, Georgia, invested $2.8 million to create a mobile farmers' market paired with a wellness clinic. The county also used a grant from a regional commission to provide meals to senior citizens who had been placed on a waitlist for food assistance.[45]

Coordinating Emergency Food

Some local governments work to ensure coordinated procurement and distribution of emergency food. They organize food reclamation, streamline service provisions, and connect food resources. The City of Portland and Multnomah County, Oregon, streamlined emergency food provision through a Schools Uniting Neighborhoods (SUN) Service System. The city and county partner with local school districts, nonprofits, and businesses to coordinate an integrated set of services for youth, families, and community members through a network of SUN schools across multiple county school districts. The SUN network partners with the Oregon Food Bank to provide 3–5 days of nutritious food to families for weekend, evening, vacation, and other non-school days when food is hard to come by. Once a month, the SUN Community School Free Food Markets distribute fruits and vegetables, also in partnership with Oregon Food Bank.[46]

BALTIMORE INITIATIVE ADDRESSES FOOD IN ALL POLICIES

Baltimore, Maryland, takes a comprehensive "Food in All Policies" approach to address economic, environmental, and health disparities in neighborhoods with high food insecurity. Adapting the "Health in All Policies" framework advanced by the American Public Health Association and the CDC, the city created the Baltimore Food Policy Initiative (BFPI) in 2009. The initiative takes a systems approach to integrate food across government agencies, strategic planning, and policy co-creation with affected community residents. It includes three pillars:

- Interagency Collaboration;
- Food Policy Action Coalition; and
- Resident Food Equity Advisors.

BFPI brings together multiple perspectives to find policy solutions. These range from modifying practices within organizations and institutions to changing city regulations to advocacy. Lead agencies include the city's Department of Planning, Health Department, Office of Sustainability, and the Baltimore Development Corporation (BDC). It addresses acute food insecurity while building long-term food resilience and advancing equity, seeking to right power imbalances and strengthen the local food economy. Toward these ends, it proposes agricultural and land use policies, including making suitable city-owned land available for food production and developing pathways to land ownership, promoting sustainable agriculture, and supporting growers to create financially viable urban agriculture. It has supported zoning changes to exempt hoop houses and allow poultry, small livestock and bees, food truck legislation, healthy vending machine procurement, and policies to support school wellness, nutrition, and physical activity.

BFPI created a Food Desert Retail Incentive Area concept. The city now gives property tax credits to supermarkets that locate or renovate in specified incentive areas and meet certain requirements for the amount of healthy food they provide. Implemented by the Baltimore Development Corporation, the tax credit also supports small loans for stores in underserved neighborhoods to invest in equipment and infrastructure, and funding for the Maryland Fresh Food Financing Initiative. Further, working with leadership of the Baltimore City Managerial and Professional Society (MAPS) and the Labor Commissioner, BFFI allows CSAs as an approved use of the existing Health Reimbursement Policy so MAPS employees can be reimbursed if they participate in a CSA.

BFPI also has advocated for changes to SNAP. State-level efforts included support for bills to provide additional SNAP benefits to children during summer and winter breaks, to extend the SNAP Issuance period from 10 to 20 days to smooth out retail cycles, and to provide a dollar-for-dollar match for purchases of fresh produce made using federal food assistance at participating farmers markets throughout the state. During Covid, it advocated for Maryland to expedite Online SNAP implementation in Baltimore City to make it possible for more to shop for groceries at home. It also advocated to HFFI to include underserved cities that fall outside of the federal food desert definition.[47]

Discussion Questions

1. What kinds of issues affect food security and health disparities in your community?
2. What policies or programs do you have to address them?
3. How are they working?
4. What more—if anything—is needed?

Notes

1. Kate Fitzgerald, Anne Palmer, and Karen Banks, "Understanding the SNAP Program for Food Policy Councils," *Johns Hopkins Center for a Livable Future*, n.d.
2. Rachel Mackey, *Supplemental Nutrition Assistance Program (SNAP) Reauthorization and Appropriations* [Policy Brief] (Washington, DC: National Association of Counties, February 1, 2023).
3. For an updated list, visit the USDA FNS website, "Farmers' Markets Accepting SNAP Benefits."
4. Center on Budget and Policy Priorities, *A Quick Guide to SNAP Eligibility and Benefits* (Washington, DC: Center on Budget and Policy Priorities, March 2023).
5. The Uconn Rudd Center for Food Policy and Health has tracked hundreds of local policies passed since 2010 on issues ranging from obesity and diet-related diseases to food access and assistance, package labeling, and food and beverage taxes. Visit their website under Research, Food Policy and Environment, Local Policy.
6. Johns Hopkins University Center for a Livable Future has a directory of food policies councils.
7. Rachel Santo et al., "Pivoting Policy, Programs, and Partnerships: Food Policy Councils Response to the Crises of 2020," *Johns Hopkins Center for a Livable Future*, 2021.
8. Sarasota County Florida, "The Sarasota County Comprehensive Plan: A Planning Tool for the Future of Sarasota County. Volume 1: Goals, Objectives, & Policies," 2016.
9. Arizona Food Systems Network, "Arizona Statewide Food Action Plan 2022–2024," 2022.
10. USDA ERS Key Statistics & Graphics, "Food Security Status of U.S. Households in 2021."
11. Kelly M. Bower et al., "The Intersection of Neighborhood Racial Segregation, Poverty, and Urbanicity and Its Impact on Food Store Availability in the United States," *Preventive Medicine* 58 (January 2014): 33–39.
12. Desiree Sideroff, "Getting to Grocery: Tools for Attracting Healthy Food Retail to Underserved Neighborhoods," *Change Lab Solutions*, 2012.
13. Michael E. Porter, "New Strategies for Inner-City Economic Development," *Economic Development Quarterly* 11, no. 1 (February 1, 1997): 11–27.
14. USDA Food and Nutrition Service's Supplemental Nutrition Assistance Program, "Healthy Corner Stores: Making Corner Stores Healthier Places to Shop," USDA, 2016.
15. The City and County of San Francisco, "The Healthy Food Retailer Ordinance," No. 193-13, Chapter 59, American Legal Publishing.
16. San Francisco Administrative Code SEC. 59.3. DEFINITIONS.
17. Massachusetts Department of Public Health, Bureau of Community Health and Prevention, "Mass in Motion."
18. USDA Food and Nutrition Service's Supplemental Nutrition Assistance Program, "Healthy Corner Stores: Making Corner Stores Healthier Places to Shop," USDA, 2016.
19. U.S. Census Bureau, SNAP Benefits Recipients in Philadelphia County/City, PA [CBR42101PAA647NCEN], Federal Reserve Bank of St. Louis.
20. The Food Trust, "Healthier Corner Stores: Positive Impacts and Profitable Changes," *The Food Trust*, 2014. Information also provided by Nyssa Entrekin, Caroline Harries,

and Julia Koprak in an interview on January 6, 2023, and a series of emails between January 3, 2023, and March 3, 2023.
21. Data supplied by Caroline Harries of the Food Trust on February 23, 2023.
22. New Orleans Fresh Food Retailer Initiative. Information Sheet.
23. Change Lab Solutions, "Licensing for Lettuce: A Guide to the Model Licensing Ordinance for Healthy Food Retailers," 2013.
24. City of Minneapolis. Minneapolis Code of Ordinances. Title 10. Chapter 203. Grocery Stores. "Staple Foods Ordinance Fact Sheet."
25. Amending Title 10, Chapter 203 of the Minneapolis Code of Ordinances relating to Food Code: Grocery Stores.
26. City of Minneapolis website, Government Programs and Initiatives, Healthy Living, Healthy Eating, "Staple Foods Ordinance."
27. ChangeLab Solutions, "Healthy Mobile Vending Policies: A Win-Win for Vendors and Childhood Obesity Prevention Advocates," *The National Policy & Legal Analysis Network to Prevent Childhood Obesity, ChangeLab Solutions*, 2013.
28. City of Kansas City, Missouri Parks and Recreation Department website, "Mobile Vending Policy."
29. Arcadia Center for Sustainable Food and Agriculture website, Our Programs, "Mobile Market."
30. David Webber et al., "The Massachusetts Farmers' Market Coupon Program for Low Income Elders," *American Journal of Health Promotion*, no. 4 (March 1, 1995): 251–53.
31. Fair Food Network, "Double Up Food Bucks 2021 Annual Impact Report."
32. Massachusetts Healthy Incentives Program website, "Massachusetts Healthy Incentives Program (HIP) Fact Sheet."
33. Hunger Free Oklahoma website, "The Work Continues into 2023," Hunger Free Oklahoma Update, January 25, 2023.
34. Fresh Bucks website, "Produce Prescription (RX) Program, Marion County Public Health Department."
35. Community Outreach & Patient Empowerment Program, Food Access, "Fruit & Vegetable Prescription Program," 2022.
36. Morgan Marshall, "NC State Extension Local Food Program Team Annual Report 2021–2022." *NC State Extension Local Food Program*, 2022.
37. School of Public Health. Harvard Healthy Eating Plate. Harvard University.
38. Health Cleveland website, "Nutrition Guidelines."
39. Navaho Epidemiology Center, "Understanding the Healthy Diné Nation Act of 2014," *Navaho Nation*, 2019.
40. City of Seattle website, "Sweetened Beverage Tax."
41. Burlington School Project website.
42. D.C. Health Schools Act brochure.
43. Feeding America, "Charitable Food Assistance Participation in 2021."
44. Seattle Department of Neighborhoods. Food Equity Fund.
45. Healthy Food Project. Municipal Policies to Support Food Access During Emergencies.
46. Multnomah County, Department of County Human Services, SUN Service System webpage.
47. City of Baltimore Department of Planning. Food Policy webpage.

8
EVOLVING ISSUES

Chapter Summary

This chapter explores three issues that affect food production but have not received much planning attention: farmland access, solar energy siting, and commercial cannabis cultivation. While policy responses are evolving, these issues have not yet been integrated into food systems planning. This chapter is a step toward providing some guidance. It concludes with some thoughts about where we go from here.

***

Chapter 3 discussed the need to plan for food systems as *systems*—not individual sectors. Systems thinking can help communities manage competing land uses by balancing priorities for agricultural vitality, food security, renewable energy, and economic development.

Access to arable and affordable land is a significant barrier to domestic food production, especially in the urban-influenced areas responsible for producing most of our dairy, poultry, and specialty crops. Land prices have reached record highs, and there is a very tight supply of suitable farmland to rent or to buy, making it hard for a new generation to enter agriculture and for small and midsized farms to expand. State and local governments are stepping up efforts to increase access to farmland, often for local food production. At the same time, states are passing Unified Partition of Heirs Property Acts to protect heirs' property owners from forced partition sales and, building on the federal example, to restore ownership of tribal lands.

Exacerbating the land access problem, more than half the states have set goals for reducing GHG emissions and dramatically increasing generation of renewable energy, especially solar. This is an important step toward environmental sustainability and resiliency. But if it happens in a vacuum without vision and planning, it will come at the expense of our farmland and food systems. Ambitious renewable energy targets require dramatic increases in the number and scale of utility solar installations, largely planned for rural lands. While the U.S. Department of Energy reports that solar needs could be met with the equivalent of less than 10 percent of potentially suitable disturbed lands,[1] without planning and policy intervention, most solar development will occur on farmland because it is cheaper and easier to develop than disturbed or marginal lands.

Finally, now legalized in some form in most states, cannabis cultivation also is competing for land and challenging communities with siting considerations. Most communities are unprepared for a green rush and its added pressure on the land base. Going forward, these issues will command more attention, with solar siting and cannabis production especially affecting rural communities.

Access to Farmland

Nearly 40 percent of all farmers and ranchers are aged 65 and older, four times as many as those under age 35[2] (see Figure 8.1). Including nonoperator landlords, seniors own more than 40 percent of American farmland.[3] Many are aging in place, which contributes to a tight agricultural land supply: Very little acreage is available to rent or to buy, especially on the open market.[4] Competition for land is fierce given a farm consolidation, rapidly appreciating land values, and a diminishing supply due to farmland conversion. These factors align to favor large farms and established operators.[5]

In the past, farmland mostly transferred from generation to generation within a farm family through cradle, alter, or grave. Today most producers do not inherit their land.[6] Even when families want to pass down their land, interpersonal dynamics and issues like fractionated ownership and heirs property rights complicate transfers between relatives. With little land

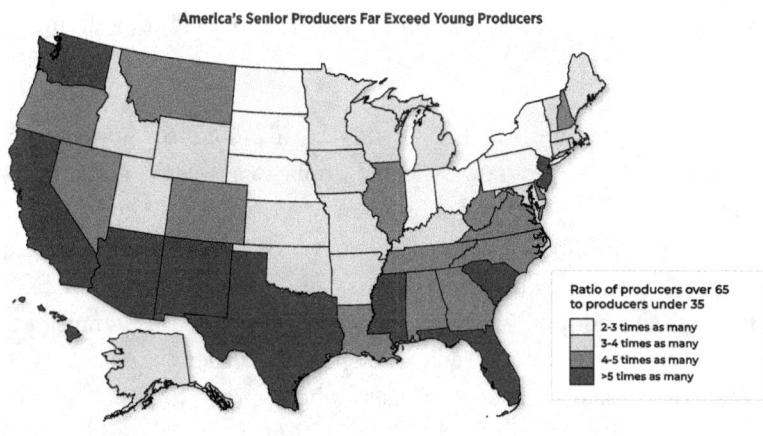

Figure 8.1 States average over four times as many senior producers as young producers—a stark contrast to the general workforce, where more than six times as many people under age 35 are employed than people over 65.

available to rent or to buy, and land values at an all-time high, land access is a major challenge, especially for a diverse new generation of young and beginning farmers entering the field.[7]

Several states and some communities have addressed these issues through tax and other policy incentives, leasing land, and supporting LandLink programs. Others have passed Unified Partition Heirs Property Acts to address land dispossession through forced sales. And a few are working with tribes to restore consolidated ownership of fractionated tribal lands.

Land Access Policy Incentives

Nebraska passed the first Beginning Farmer Tax Credit Act in 1999. Since then, a handful of states have developed financial incentive programs to help young and beginning farmers gain access to land.[8] *Beginning Farmer Tax Credit* programs give agricultural landowners a credit on their state income taxes in exchange for renting and/or selling farmland and other agricultural assets to beginning farmers. Nebraska's law gives asset owners state income tax credits for entering into three-year rental agreements with beginning farmers. Assets include land, livestock, buildings, and machinery. Iowa's program is similar, while Kentucky only incentivizes sales. Minnesota, Pennsylvania, and Ohio offer credits for both sales and rentals. *Farm Purchase and Protection programs* use a PACE framework mostly to help young and/or beginning farmers buy farmland. They use various mechanisms:

- Maryland's program is a fast-moving farmland conservation easement option purchase program. Selected applicants receive down-payment funds to buy farmland and subsequently protect it (see box).
- Delaware provides a zero-interest loan in exchange for an easement on farmland purchased through the program.
- Pennsylvania adopted an exemption from the state's realty transfer tax if a protected farm is conveyed to a beginning farmer.
- Washington state created a tax-exempt bond program to help beginners and historically underserved farmers and ranchers acquire farmland at lower interest rates.
- Rhode Island has a *Buy Protect Sell* program, which purchases farmland from willing landowners at fair-market value, restricts it with an agricultural conservation easement, and then sells it to a qualified farmer at restricted agricultural value.

MARYLAND'S NEXT GEN PROGRAM SUPPORTS FIRST-TIME FARM BUYERS

Figure 8.2 Maryland's Next Gen program helps young and beginning farmers acquire and protect farmland.

Source: USDA/FPAC. Photo by Preston Keres.

Maryland's Next Gen Farmland Acquisition Program grew out of recommendations from the 2006 Statewide Plan for Agricultural Policy and Resource Management. It addresses two challenges: 1. helping young and beginning farmers acquire farmland and 2. protecting that land from future development. Enacted as part of the Maryland Agricultural Stewardship Act of 2006, due to the Great Recession, funds were not allocated until 2018.

Next Gen is administered by the Maryland Agricultural & Resource Based-Industry Development Corporation (MARBIDCO), a state economic development authority. MARBIDCO provides down-payment funding to qualified first-time farm seekers who then have several years to sell an easement to a county or state PACE program. Once the easement is sold, they repay MARBIDCO, plus a modest administrative fee. If they cannot sell an easement within the time frame, the option is exercised and the easement assigned to a county program or a private land trust. MARBIDCO also has a Small Acreage Next Generation Program to help qualified young or beginning farmers purchase small farmland properties between ten and 49 acres that are ineligible for the original Next Gen Program.

Land Banks

Land banks are governmental, quasi-governmental, or nonprofit authorities created to acquire vacant, abandoned, and foreclosed properties. Once acquired, they maintain these properties until they can be transferred to other entities which return them to productive use. Created under state enabling legislation and enacted by local ordinances, their functions and structures vary widely, but generally they authorize local governments to acquire properties, eliminate tax liens, provide clear title to foreclosed properties, and foster community development. Land banks are different from land trusts, which hold properties in perpetuity for a community purpose, such as affordable housing, environmental conservation, or farmland protection.

Land banks can play an important role in redeveloping degraded areas. They also can be used to facilitate land access for local food production, especially in urban areas. For example, the Kansas City, Kansas land bank is authorized to lease land for community gardens, urban farms, and farmers' markets. The Ingham County Land Bank in Michigan has a garden program which leases land for larger-scale urban agriculture as well as for household and community gardens. And the Hartford Connecticut Land Bank partnered with a local nonprofit to create an urban farming initiative in a federally designated Promise Zone to produce fresh produce while finding new uses for vacant lots.

LandLink Programs

LandLink—or FarmLink—programs facilitate farm transfer by connecting retiring landowners with farmland seekers.[9] Most programs are operated by NGOs, but ten states implement and/or invest public dollars in these programs. Eight are managed by Councils of Government or Cooperative Extension; only three have state authority.[10]

LandLink programs have websites with resources, including sample leases and searchable databases, to facilitate connections. Successful programs have personnel dedicated to help with transactions and provide technical assistance. For example, staff in Connecticut's program offer site assessments, help develop leases and purchase and sale agreements, and provide advice on issues ranging from production to protecting land with an agricultural easement. Along with access to land and high-quality land tenure, California FarmLink supports farmers and ranchers with access to financing and business education. New York's program is a partnership between the state and American Farmland Trust. Its farmland finder website includes listings of farm properties and farmers seeking land, on-farm jobs, events, and other resources. The program also supports a statewide network of partner organizations with dedicated staff who provide training and support for farmers and landowners.

Leasing Public Lands

Many public agencies lease land to farmers and ranchers. They include natural resource and parks departments to BLM and the Department of Defense: BLM alone manages grazing on 155 million acres of public land.[11] States also make trust lands available—lands Congress gifted mostly to western states to support public institutions, including agricultural colleges.

While most states allow leasing of state-owned lands, few prioritize agriculture. Nine allow it on land managed as wildlife habitat with restrictions on activities. For example, in Iowa's lease to beginning farmers program, the state Department of Natural Resources (DNR) leases wildlife habitat lands to certified beginning farmers. Leases run for up to seven years and may be renewed, with preference given to beginners who have not yet participated. DNR may require conservation systems and adoption of generally accepted farming or soil conservation practices as long as they are compatible with resource management and outdoor recreation policies.[12]

Only five states give farms and ranches preference over other kinds of land use. Hawaii's Agricultural Parks Program leases land to small, diversified farmers to improve local agricultural viability. To ensure the land is conserved for long-term production, lessees must be established farmers or qualified beginners. California passed legislation to create Urban Agricultural Incentive Zones. It authorizes local governments to contract with landowners to restrict their land to small-scale production in exchange for agricultural tax rates. Contracts cover vacant, unimproved, or otherwise blighted lands, must be for a term of no less than five years, and restrict property that is at least 0.10 acres and no more than three acres in size.

Local governments also lease land for food and farm production. They address issues such as allowing public access, investments in soil health and farm infrastructure, water access, equipment storage, manure spreading, and so on. Some manage extensive public land resources. Boulder County, Colorado, leases 25,000 acres to operations ranging from small market gardens to large commodity farms, including 7,000 acres for grazing livestock. Over 90 percent of crops grown on its land end up in the food system. Leases are both cash payments and crop shares where the farmer and the county split portions of the harvest.[13] Other communities have smaller programs that often prioritize food production. Lawrence, Kansas's Common Ground program leases under-utilized properties to residents for free or at a very low cost for community gardening and urban agriculture. Madison, Wisconsin, has ordinances for community gardens and to encourage edible landscaping on city-owned land.

Urban Agriculture Tax Credits

Some state and local governments offer tax credits and other incentives for urban farms. California was one of the first, passing legislation in 2013 to create urban agriculture incentive zones. Assembly Bill No. 551 authorized

Figure 8.3 The City of Ontario, California, received a $1 million grant from the Kaiser Permanente Healthy Eating Active Living (HEAL) Zone initiative. Some of this went to scale up food production at Huerta del Valle, a four-acre organic Community Supported Garden and Farm. The city supported the project beyond the grant by providing a vacant piece of land next to a residential park and community center.
Source: USDA. Photo by Lance Cheung.

cities and counties to enter into ten-year contracts with landowners to dedicate vacant, unimproved, or otherwise blighted parcels of at least 0.10 acres to small-scale agriculture.[14] The following year, Maryland updated its tax code to allow for tax credits for urban agriculture.[15] More recently, Missouri approved tax credits to establish or improve urban farms and community gardens that produce food for public distribution. Taxpayers may receive state income tax credits for up to 50 percent of eligible expenses up to $5,000. The credit is not refundable, but any excess can be carried forward and applied to the succeeding three years.[16]

To encourage continuous agricultural use and maintenance of otherwise vacant land, Baltimore gives farmers 90 percent off their property taxes if their land is used for urban agriculture for five years and produces a minimum threshold of value. San Francisco's program requires applicants to demonstrate they will benefit the larger community through produce distribution and/or sales, open house days, educational tours, or other public programs. Eligibility is limited to property owners located in zones that allow agricultural uses. Parcels must be vacant land between 0.10 and three acres in size and may only include structures that are accessory to the agricultural activity, including but not limited to toolsheds, greenhouses, produce stands, or educational space.[17] Washington, D.C., offers up to a 90 percent property tax abatement for properties actively used as urban farms up to a maximum annual abatement of $20,000.[18]

Heirs Property and Fractionated Lands

Properties transferred to heirs without a will are called tenancy-in-common—or heirs' property. When a landowner dies without a will, their property passes down to heirs as undivided shares—or fractional interests. Over generations, legal title to this land becomes increasingly ambiguous with dozens—sometimes hundreds or thousands—of potential heirs. Predominant among Black/African Americans in the South, heirs' property affects other populations, including Appalachian and Hispanic producers, and tribes as fractionated lands.[19]

Without clear title, heirs may use their family property but have trouble borrowing money and until recently could not participate in federal farm programs and disaster relief. (The 2018 Farm Bill remedied this by authorizing alternative documentation so heirs' property owners could establish a farm number to become eligible for USDA programs.) Further, since any one heir can sell their share, heirs' property owners are vulnerable to third-party partition sales. Without due process, partition sales have been exploited by real estate speculators who acquire a small share to force the sale of an entire property—often against the wishes of the majority owners. They also have paid delinquent taxes and obtained a deed to a property without the owners' knowledge or consent.[20] Forced partition sales have been a major contributor to the $326 billion estimate of Black land loss over the course of the 20th century.[21]

The Uniform Partition of Heirs Property Act (UPHPA) addresses dispossession of land through forced sale. It creates a fairer—if still complex—partition process, protecting the right of co-inheritors to sell their interests while providing other family members with due process to protect their interests. Protections include notice, appraisal, and right of first refusal. Since 2010, 23 states have passed UPHPA statutes, with several more pending.[22]

Land Buy Back programs restore ownership of tribal lands. The 1887 General Allotment (Dawes) Act divided up reservation land and allotted it to individual tribal members. It allowed 60 million acres of "surplus" lands to be sold or transferred to white settlers and led to the loss of another 30 million acres through forced sales and other takings. These lands remained within reservation boundaries but outside of tribal ownership and control.[23]

Similar to heirs' property, when the allottee of the original Dawes Act died, ownership was distributed among heirs as undivided interests. Through many generations, this led to highly fractionated ownership of tribal lands. The U.S. government holds some of these lands in trust, with restrictions on use and disposition of the land. Other land is "restricted," meaning a tribe or tribal member holds title but the federal government still can impose restrictions on its use and/or disposition. These lands are eligible for buy-back purchases.

A federal Land Buy-Back Program for Tribal Nations provided a $1.9 billion Trust Land Consolidation Fund. Authorized from 2012 to 2022, it was used to purchase fractional interests and restore the consolidated interests to tribal ownership. It was available to landowners who owned mostly small, fractional interests.[24] Tribes continue to work with NGOs and state governments to reclaim lands in other ways. For example, in 2020 the Esselen Tribe bought back 1,200 acres of ancestral land near Big Sur, California, with help from a $4.5 million grant from the California Natural Resource Agency and an Oregon-based environmental group. So far, these efforts have focused on land used for things other than farm and food production, but tribes are building capacity to engage in many types of real estate transactions. The Indian Land Capital Company is a Native-owned, Certified Native Community Development Financial Institution (CDFI) that provides alternative loan options to Native Nations for tribal land acquisition and economic development projects, including to help tribes consolidate undivided interests in fractionated land.

Solar Energy Siting

Solar energy is a key to energy independence and to reducing GHG emissions. But without proper planning and land use regulations, it develops farmland and is a threat to agricultural viability. While pursuing renewable energy goals, communities must balance the impacts of solar energy development on food production.

The U.S. Department of Energy (DOE) posits that with supportive policies, aggressive cost reductions, and large-scale electrification, solar could account for as much as 40 percent of the nation's electricity supply by 2035 and 45 percent by 2050. To achieve this, solar deployment must grow by four times its current rate—with 90 percent occurring on rural lands.[25]

DOE projects that solar needs could be met with the equivalent of less than 10 percent of potentially suitable disturbed lands.[26] It even has a guide to help local governments evaluate opportunities to site community-scale solar projects on landfills and other contaminated sites.[27] But developing brownfields, landfills, and other disturbed lands is costly, complicated, and time consuming. On the other hand, developing farmland is cheap, easy, and fast because the land is flat, cleared, unshaded, and drained. It also is more likely to have high solar radiation and to be close to existing transmission infrastructure.[28]

Rural communities have fewer land use regulations to constrain or direct development. AFT modeling suggests that without intervention, by 2040, 83 percent of solar development will take place on farmland, with nearly

Figure 8.4 Without better planning, by 2040 as much as 40 percent of solar development could occur on our most productive cropland.
Source: iStock photo by freedom-naru.

half on our most productive cropland.[29] With the most extensive solar development projected to occur in three of the largest agricultural states (California, Florida, and Texas), what will be the impact on domestic food production?

In theory, if a solar array is installed to be sensitive to future agricultural use, the conversion is temporary. It would require construction, operation, and decommission standards to protect soil health and productivity. But leases often are for 35 years or more, and many give an option to renew with a right of first refusal for the developer. Given the cost of infrastructure investment and the fact that a good site for solar today will be a good site for solar tomorrow, it is unlikely farmland developed for utility solar arrays will be restored to full productivity.

Beyond the actual conversion of farmland, developing large expanses of farmland for utility solar inflates land values and threatens agricultural viability. It displaces farmers who currently rent land and increases barriers to land access, especially for young, beginning, and historically disadvantaged producers. It also can destroy pollinator habitat. Some communities have developed zoning ordinances to address these threats. Scott County, Iowa, with its rich agricultural resources and strong land use policies, created a special zoning district to achieve smart solar siting (see Chapter 5), but most have not.

Incentivizing Solar on Disturbed and Developed Sites

With effective planning, communities can protect farmland and encourage solar development at the same time. Along with brownfields and other disturbed sites like parking lots, they can incentivize solar development on warehouses, large agricultural structures, non-farmable ditches, and so on, advancing renewable energy while protecting valuable agricultural land.

Parking lots offer a huge, and as yet barely tapped, potential for solar development. The U.S. Geological Survey calculated the proportion of each county's land that is covered by parking lots. It estimated they cover 13,778 square miles—0.47 percent of the nation's contiguous land area.[30] Other studies have found that parking takes up about one-third of land area in American cities alone.[31] Using parking lots for solar arrays offers multiple benefits: shade in summer, shelter from inclement weather, and charging stations for electric vehicles. Recognizing this potential, in 2022 France approved a measure to require large parking lots to include solar canopies.

EVOLVING ISSUES

Figure 8.5 Heidel Hollow Farms is a 1,800-acre operation in Germansville, PA. With help from USDA, they installed an 896-panel solar array on diversion ditches that could not be farmed, providing up to 70 pecent of the farm's electrical needs.

Source: USDA. Photo by Lance Cheung.

The law applies to all parking lots with surface areas more than 1,500 square meters (80 spaces or more) and will be phased in over a five-year period.[32]

New York State's Build Ready program partners with local communities to identify and advance under-utilized land for renewable energy projects, including parking lots. To make these sites attractive, the state's Energy Research and Development Authority (NYSERDA) offers a de-risked package for developers to construct and operate projects, including contracts for renewable energy payments. NYSERDA takes difficult sites and makes them "build-ready," reducing barriers to developing solar on brownfields, landfills, parking lots, commercial, industrial, and other underutilized sites.

Mitigation Fees

In cases where alternatives to developing farmland are not available, state and local governments can impose mitigation fees. These can be used to fund PACE programs. New York requires a mitigation fee on solar projects that receive state solar incentives in designated agricultural districts. Developers pay higher fees for more productive soils. Fees are paid to the state's Agricultural and Farmland Viability Protection Fund to support farmland protection projects across the state. In California, mitigation is a common requirement for projects that convert farmland to non-agricultural uses, including solar. The California Council of Land Trusts offers guidance to communities to develop and refine their programs.[33] Beyond mitigation, policies can incentivize topsoil retention, which is mutually beneficial for agriculture and pollinators.[34] Seven states have passed legislation to allow solar projects to claim that they are pollinator friendly.[35]

Community Guidance

Some states issue solar siting guidance to communities. Rhode Island's Solar Energy Systems Ordinance Template was developed to help municipalities address solar energy systems in their zoning ordinances. It provides definitions and addresses a variety of solar development forms so communities can decide what is appropriate and compatible with their comprehensive plans. And it offers guidance on where to allow solar arrays for primary or accessory use in each zoning district, along with development standards, such as buffers, height, and storm water control, related to scale and land use context, including development of farmland.[36]

NRCS's *Land Evaluation and Site Assessment (LESA)* model is a useful tool for identifying disturbed and marginal lands for utility-scale solar installations. It provides a technical framework to rank land parcels based on resource evaluation and site considerations. In evaluation, soils are rated and grouped based on their suitability for a specific agricultural use. Site assessment assigns a range of values to landscape factors related to development pressure, non-soil factors related to a site's agricultural use, and other public values. Each factor is assigned a range of values based on local needs and objectives to provide a sound and consistent basis for making land use decisions. Other tools include financial incentives and providing guidance on "smart solar" siting.

Taking a landscape approach and prioritizing land uses also help to avoid conflicts with solar development. Beyond using tools like LESA, communities benefit from engaging stakeholders in a prioritization process. For example, a team of planners and conservationists in California's San Joaquin Valley facilitated a six-month stakeholder-led process and successfully identified land with the least amount of conflict to develop for solar installations (see box).

STAKEHOLDERS IDENTIFY LEAST-CONFLICT LANDS FOR SOLAR DEVELOPMENT

California's San Joaquin Valley (the Valley) is vital to the nation's food supply. Responsible for hundreds of crop and livestock products, the breadbasket of the nation is one of the world's most productive agricultural regions. It also is one of the fastest growing and contains some of the state's most threatened natural habitats.

More than half of the Valley's high-quality farmland is threatened by low-density housing development.[37] With high solar insolation and a temperate climate, it is under increased development pressure to achieve California's aggressive renewable energy goals. With hundreds of proposed projects, each averaging about 500 acres in size, appropriate siting is essential to the future of its agriculture and food production.

Identifying suitable locations for solar installations was controversial. Conservation organizations advocated for "smart from the start" planning to encourage developers to locate solar projects near existing transmission corridors and in areas with low environmental value[38] while farm organizations pushed to avoid lands in agricultural production. With input from the Governor's Office of Planning and Research, a project team of planners and conservationists initiated a stakeholder-led process to identify least-conflict lands for solar photovoltaic development. The team included Conservation Biology Institute, Berkeley Law Center for Law, Energy and the Environment, and Terrell Watt Planning Associates. Diverse stakeholders included farmers and ranchers, environmentalists, and the solar industry.

The team used advanced mapping software to generate a series of maps, created a *San Joaquin Valley Data Basin Gateway* website[39] and used these to work with each stakeholder group to identify their highest- and lowest-priority areas. After all the groups met, the team combined the results to identify 470,000 acres of least-conflict land, about 5 percent of the study area. This provided a credible snapshot to inform future planning and policy efforts to balance renewable energy with agricultural and conservation interests. The Gateway is still live and includes a valuable data library with hundreds of relevant datasets to inform future planning.

Smart Solar[SM] Siting

Smart Solar[SM] addresses competing needs to contain solar development on high-quality farmland. The concept, as developed by AFT, borrows from smart growth principles to guide solar development onto land where it has the least negative impact on soils and farm production and encourages *agrivoltaic* projects to support compatible farming systems.

Still largely untested, agrivoltaic projects are considered *dual use* for solar and agriculture. Under the right conditions, they can work well on marginal lands suitable for grazing poultry and small livestock (see Figure 8.6), although they still limit most agricultural production. Smart Solar[SM] projects have three goals:

1. Accelerate solar energy development;
2. Strengthen farm viability; and
3. Retain high-quality farmland.

To achieve these goals, projects prioritize development on disturbed lands and within the built environment and avoid high-quality agricultural soils.

Figure 8.6 Sheep graze under solar panels in a Smart Solar[SM] project.
Source: Photo courtesy of NCAT's Solar Clearinghouse.

If a project must involve farmland, it directs siting to marginal soils and modifies solar arrays to be as compatible as possible with farming on the same land. Arrays are planned, developed, and ultimately decommissioned to allow for continued agricultural use and to limit adverse impacts, including soil disruption.

New techniques are being tested to install poles with minimal soil disturbance, adjust panel heights, and create wide enough space between panels to accommodate equipment, grazing, and shaded vegetable production. Only modest amounts of land are needed for posts, relay boxes, or other infrastructure, to keep most of the land available for production.

Agrivoltaic approaches warrant continued experimentation and proof-of-concept research in different climates and with different farming systems. More research is needed to identify the types and scales of cropping and livestock systems that are compatible for dual use, and their impacts on farm yields and regional economies.

Cannabis Cultivation

The cannabis plant includes hemp and marijuana. Legally, the difference between them is the amount of tetrahydrocannabinol content (THC), which causes the "high" associated with its use. Laws limit hemp's THC content to 0.3 percent or less by dry weight; marijuana is anything more. Both are used medicinally, primarily non-intoxicating cannabidiol (CBD). Hemp generally contains more CBD and marijuana more THC.

Humans have cultivated hemp for millennia for all kinds of products. In the 17th century, it was considered an important fiber crop in the New England colonies. Peak U.S. production occurred in the 19th century but decreased in the early 20th century as other fiber crops and synthetic fibers led to declining demand and as Congress began to regulate marijuana. Hemp production increased during World War II to produce raw materials for rope and canvas, but quickly declined until 1970, when the Controlled Substances Act listed "marihuana" as an illegal Schedule 1 narcotic, making no distinction between the drug and hemp.

The 2018 Farm Bill legalized hemp production. While four states still ban it, it is now legal under federal law and eligible for many USDA programs. Since 1996, when California passed the first statewide medical marijuana law, 37 states, three territories, and the District of Columbia have followed

Figure 8.7 Jeff Garland (right) gives Indiana NRCS district conservationist Lee Schnell a tour of high tunnels on Papa G's Organic Hemp Farm in Crawford County, IN.
Source: NRCS. Photo by Brandon O'Connor.

suit. Colorado passed the first recreational law in 2014. By 2023, 23 states, two territories, and the District of Columbia have enacted measures to regulate cannabis for adult non-medical use.[40]

Marijuana remains illegal under federal law. In states where it is legal, it must be grown within the state where it is sold and cannot be transported across state lines. Since its acreage is not tracked, it is unclear how much impact it has on land use. But with farmland already in short supply, communities are ill prepared for a new green rush. Guidance is scarce, mostly focused on use, distribution, and sales but not on land use. Where hemp generally is allowed as an agricultural use, only some states classify marijuana as agriculture; others consider it a recreational land use. Contradictory laws and classifications add to the confusion. However it is classified, cannabis cultivation is similar to other agricultural crops and communities are developing zoning and other land use regulations to address it. These often vary based on whether production takes place indoors, outdoors, or in greenhouses.

Table 8.1 Considerations for allowing cannabis cultivation.

Indoor Production: Large scale in warehouses and similar facilities	Provides most environmental control; Reduces use of pesticides and other inputs; Largest environmental impact; Requires substantial energy use for lighting, climate control, and automated equipment; Highest potential for light pollution; Requires good ventilation to minimize strong odors from leaving the premises; and Best suited to urban environments and industrial or manufacturing zones.
Greenhouse Production: Small scale in greenhouses or hoop houses	Lower environmental impacts than large warehouse facilities; Reduces use of pesticides and other inputs; Requires ventilation, climate control, and other equipment; Can be managed to minimize noise, light, and odors; and Most compatible with other land uses in a variety of settings.
Outdoor Production	Requires rich soil with organic matter, good drainage, and ample water supplies; Cheaper than indoor alternatives; Creates a smaller carbon footprint; Produces strong odors at some times of the year; Often requires irrigation, pesticides, and other inputs; Security can be a problem; and Best suited to rural areas with distance from residential neighbors.

Some communities allow production in certain areas, for example, indoor production in industrial zones, outdoor in agricultural. Others require a conditional use permit and regulate issues like light pollution, water use, or odors. Requiring site plan review or an Environmental Impact

Statement helps to ensure operations meet local standards. Siting considerations include where production may occur, minimum lot or parcel size, hours of operation, traffic or road use, parking and loading, landscaping and screening, and environmental impacts. Common considerations include:

Buffers and Setbacks: Most marijuana regulations require setbacks from places where children congregate, like schools, childcare centers, and playgrounds. Setbacks generally range from 500 to 1,000 feet. Setbacks from residences and other land uses may be as little as 25 feet but more often between 50 and 200 feet. For outdoor cultivation, fences and other site screenings can reduce vandalism and conflicts with neighbors. Vegetative buffers and buffer zones create visual and/or physical distance from other land uses. Requirements address distance from incompatible land uses, the size and type of materials and/or vegetation used for buffer strips or screening.

Canopy: Some state and local governments regulate the canopy—or surface area used to produce mature cannabis plants. Minimum lot size, setbacks, and buffers work well to regulate where production is allowed and how close an operation can be to neighboring property lines. Minimum lot sizes generally range from five to 20 acres, including setback and buffer requirements.

Energy and Light: Indoor production requires substantial energy use. Some communities address scheduling to avoid times of peak demand and create lighting standards. Lighting standards include requiring energy-efficient lighting, limiting the use of grow lights to daytime hours, or requiring "Dark Sky" compliance to meet the International Dark Sky Association's requirements for reducing waste of ambient light. IDA provides information on lighting ordinances and approved a model lighting ordinance template to help communities develop standards to reduce glare, light trespass, and skyglow.[41]

Odors: Whether for hemp or marijuana, cannabis releases strong odors during parts of the growing season. Some communities require ventilation and/or air filtration systems for indoor operations to neutralize airborne odors. For example, as part of its special permit process, the Olivette Missouri City Council requires an odor mitigation plan and for medical marijuana facilities to use odor control systems to prevent unpleasant smells from leaving the premises.[42]

Siting: Like other crops, cannabis requires arable soil, water, and light. Communities may limit production to certain areas or types of land to protect high-quality soils for food or other crop production. They also may protect cultural resources, especially if site preparation requires substantial land clearing, as this may unearth archeological resources, human remains, and sites of cultural value to tribes.

Water Use and Disposal: Cannabis is a water-intensive crop that leads to concerns about environmental management.[43] Regulations can address water use, potentially including a water management plan. To prevent depleting public water supplies and neighboring wells, some regulations require operations to have sufficient flow rates in gallons per minute or a water insurance policy or escrow account. Communities also address wastewater, including requirements on disposal methods and containment of contaminants or hazardous materials.

Concluding Thoughts

This chapter summarizes three pressing and evolving issues, but history suggests that other issues will soon emerge. The best way to prepare for what is coming is through good planning that includes agriculture and food systems as a forethought, not a footnote.

With vision and planning, communities can set in motion a transformation as profound as the one that occurred in the 20th century. State, local, and tribal governments can address farms and food in land use plans, develop stand-alone plans for food and/or agriculture, and address food systems in climate action and emergency management plans. They can step up support for farmers and ranchers, encourage a new generation, promote soil health and environmental management, and advance food and farming systems that are sustainable, resilient, and fair. Ideally, this will occur within a larger universe of smart growth and life-cycle thinking.

Food systems issues are structural—they must be addressed as systems without pitting people or sectors against each other. By lifting up civic engagement and taking a systems approach, communities can plan for renewable energy *and* protect valuable soils for food and other farm production. They can confront the historic causes of land dispossession *and* advance fair and just access to arable land. They can create buffers and setbacks to protect residents from odors and environmental conflicts, protect farmers from frivolous nuisance complaints, and so on.

Still, planning alone will not be enough. It also will take collective action and policy change to retool domestic and global food systems to achieve the diversity, flexibility, and redundancy needed to prepare for polycrisis, especially climate change. Supporting local and regional food systems is a good place to begin—not to replace but to enhance our dominant food systems. It is something we can do individually, voting with our forks as well as our votes. And it can be achieved through private markets and at all levels of government.

As we learned through Covid-19, LRFS are highly responsive: Producers pivoted quickly to adapt to disruptions and changing market conditions. They stimulated local economies, creating new businesses and jobs through food processing, distribution, and other enterprises. For LRFS to be most effective, urban and rural communities must work together to achieve economies of scale at the foodshed level, ensuring farmland is available for food production, expanding markets and infrastructure for small and mid-sized producers, and improving food access in rural communities as well as in cities.

Going forward, we must consider how the actions we take today will meet current demands without compromising future generations. We must consider how to prepare for sudden disruptions, especially their impacts on vulnerable people. And we must determine how to balance food systems actions so they reflect the full spectrum of sustainability and resiliency needs in our communities. This is no easy task. But with foresight, community engagement, and a commitment to action, it is possible. So let's begin.

Discussion Questions

1. What forces affect competition for land in your community, and what can be done to address them?
2. How would you balance competing demands for high-quality soils?
3. Does your community value local/regional food production? If so, why? If not, why not?
4. What other issues are emerging that affect planning for agriculture and food systems?

Notes

1. U.S. Department of Energy, Office of Energy Efficiency and Renewable Energy, "Solar Futures Study."
2. USDA, National Agricultural Statistics Service, 2022 Census of Agriculture – National Data. Table 52: *Selected Producer Characteristics* (Washington, DC: USDA-NASS, February 2024).
3. USDA, Economic Research Service, 2014 *Tenure, Ownership, and Transition of Agricultural Land Survey* (Washington, DC: USDA-ERS, November 2015).
4. USDA, National Agricultural Statistics Service, 2012 *Census of Agriculture Highlights: Farmland Ownership and Tenure* (Washington, DC: USDA-NASS, September 2015).
5. Christopher Burns et al., *Farmland Values, Land Ownership, and Returns to Farmland, 2000–2016, Economic Research Report* (Washington, DC: USDA Economic Research Service, February 2018).
6. Ani L. Katchova and Mary Clare Ahearn, "Dynamics of Farmland Ownership and Leasing: Implications for Young and Beginning Farmers," *Applied Economic Perspectives and Policy* 38, no. 2 (2016): 334–50.
7. USDA National Agricultural Statistics Service, "Direct Farm Sales of Food: Results from the 2020 Local Food Marketing Practices Survey," April 2022.
8. Julia Valliant and Julia Freedgood, "Land Access Policy Incentives: Emerging Approaches to Transitioning Farmland to a New Generation," *Journal of Agriculture, Food Systems, and Community Development* 9, no. 3 (May 15, 2020): 71–78.
9. Julia Valliant et al., "Fostering Farm Transfers from Farm Owners to Unrelated, New Farmers: A Qualitative Assessment of Farm Link Services," *Land Use Policy* 86 (July 1, 2019): 438–47.
10. Julia Freedgood et al., *Farms Under Threat: The State of the States* (Northampton, MA: American Farmland Trust, May 13, 2020).
11. U.S. Department of Interior, Bureau of Land Management, "Livestock Grazing on Public Lands."
12. Iowa State Legislature, "§456A.38 Lease to Beginning Farmers."
13. "Agricultural Lands on Open Space," Boulder County.
14. "AB-551 Local Government: Urban Agriculture Incentive Zones.," Pub. L. No. Assembly Bill No. 551, Chapter 406.
15. "HOUSE BILL 223: Property Tax Credit – Urban Agricultural Property – Applicability," Pub. L. No. 223.
16. "SB717 – Authorizes a Tax Credit for Urban Farms" (2022).
17. City and County of San Francisco Department of Public Health, "Application Packet for an Urban Agriculture Incentive Zone," February 20, 2015.
18. District of Columbia Department of Energy and Environment, "Urban Farm Tax Abatement Program | Doee."
19. Cassandra J. Gaither et al., "Heirs' Property and Land Fractionation: Fostering Stable Ownership to Prevent Land Loss and Abandonment," E-Gen. Tech. Rep. SRS-244. Asheville, NC: U.S. Department of Agriculture Forest Service, 2019: 1–105.
20. American Bar Association, "Restoring Hope for Heirs Property Owners: The Uniform Partition of Heirs Property Act," October 1, 2016.
21. Dania V. Francis et al., "Black Land Loss: 1920 – 1997," *AEA Papers and Proceedings* 112 (May 2022): 38–42.

22. The Uniform Law Commission Tracks UPHA legislation: "Partition of Heirs Property Act – Uniform Law Commission."
23. "Land Tenure Issues," Indian Land Tenure Foundation.
24. U.S. Department of Interior, "Land Buy-Back Program for Tribal Nations: Fractionation," February 25, 2019.
25. U.S. Department of Energy, Office of Energy Efficiency and Renewable Energy, "Solar Futures Study," 2021.
26. Ibid.
27. U.S. Environmental Protection Agency, "Community Solar: An Opportunity to Enhance Sustainable Development on Landfills and Other Contaminated Sites," December 2016.
28. Travis Grout and Jennifer Ifft, *Approaches to Balancing Solar Expansion and Farmland Preservation: A Comparison across Selected States* (Ithaca, NY: Charles H. Dyson School of Applied Economics and Management, Cornell University, March 2018).
29. Mitch Hunter et al., *Farms Under Threat 2040: Choosing an Abundant Future* (Washington, DC: American Farmland Trust, June 29, 2022.
30. James A. Falcone and Michelle A. Nott, *Estimating the Presence of Paved Surface Parking Lots in the Conterminous U.S. from Land Use Coefficients for 1974, 1982, 1992, 2002, and 2012* (U.S. Geological Survey, 2019).
31. Daniel Baldwin Hess and Jeffrey Rehler, "Parking Reform Could Reenergize Downtowns – Here's What Happened When Buffalo Changed Its Zoning Rules," *The Conversation*, June 10, 2021.
32. Projet de loi n°443, 16e legislature, adopté par le Sénat relatif à l'accélération de la production d'énergies renouvelables, November 8, 2022, www.assemblee-nation-ale.fr/dyn/16/textes/l16b0443_projet-loi#D_Article_11
33. Kate Kelly and Darla Guenzler, Editor, Dawn Van Dyke, "Conserving California's Harvest: A Model Mitigation Program and Ordinance for Local Governments," California Council of Land Trusts, 2014.
34. Leroy J. Walston et al., "Examining the Potential for Agricultural Benefits from Pollinator Habitat at Solar Facilities in the United States," *Environmental Science & Technology* 52, no. 13 (July 3, 2018): 7566–76.
35. Georgena Terry, "State Pollinator-Friendly Solar Initiatives," *Clean Energy States Alliance*, January 2020.
36. Rhode Island Office of Energy Resources and The Division of Statewide Planning, "Rhode Island Renewable Energy Guidelines: Solar Energy Systems (SES) Model Ordinance Templates, Zoning and Taxation," February 2019.
37. American Farmland Trust and Conservation Biology Institute, *San Joaquin Land and Water Strategy: Exploring the Intersection of Agricultural Land & Water Resources in California's San Joaquin Valley* (American Farmland Trust and Conservation Biology Institute, July 2018).
38. Kate Kelly and Kim Delfino, "Smart from the Start: Responsible Renewable Energy Development in the Southern San Joaquin Valley," 2012.
39. The San Joaquin Valley Gateway website is a tool to help planners and resource managers develop and implement integrated multiple-benefit solutions for long-term environmental and economic sustainability. https://sjvp.databasin.org/
40. National Conference of State Legislatures, "State Medical Cannabis Laws," Updated June 2023.

41. Dark Sky website, Resources, Guides and How-to, "Lighting Ordinances."
42. Ordinance 2760 amending the Olivette Municipal Code by enacting a new Chapter 450 Medical Marijuana Facilities, as adopted by the City Council on November 8, 2022.
43. Jennifer K. Carah et al., "High Time for Conservation: Adding the Environment to the Debate on Marijuana Liberalization," *Bioscience* 65, no. 8 (2015): 822–29.

Appendix

ORGANIZATIONAL RESOURCES

American Farmland Trust (AFT) is a national nonprofit organization that works to save the land that sustains us by protecting farmland, promoting sound farming practices, and keeping farmers on the land. See www.farmland.org

    AFT's Farmland Information Center is a clearinghouse for information about farmland protection and stewardship. It has a staffed answer service and searchable website with extensive resources to support planning activities. See www.farmlandinfo.org

American Planning Association (APA) is a national organization that elevates and unites the planning profession to help communities, their leaders, and residents anticipate and navigate change. See www.planning.org

    APA's Food Systems Division is a coalition of planners and allied professionals who advance food systems planning. The division serves as a platform for collaboration, information, and leadership to integrate food systems planning with other areas of community and regional planning for the benefit of present and future generations.

    APA's *Sustaining Places: Best Practices for Comprehensive Plans* provides a framework to turn principles into a plan and score the results, providing processes and pointers for communities to use.

ChangeLab Solutions is a nonprofit organization that uses the tools of law and policy to advance health equity. Staff partners with communities across the nation to improve health and opportunity by changing laws, policies, and systems, sharing resources, and offering practical policy tools, planning expertise, and technical assistance. See www.changelabsolutions.org/

CivicWell (formerly the Local Government Commission) supports leaders responding to the climate crisis and its impact on their communities. It is a nonprofit organization that equips, connects, and cultivates leaders working toward a more sustainable and resilient future through policy guidance, collaborative partnerships, and direct assistance. See https://civicwell.org/about-us/

Cooperative Extension is a nationwide research and education system that, among other things, supports agriculture, natural resources, health and nutrition, community development, and emergency preparedness. It operates through state land-grant universities in partnership with federal, state, and local governments, and the USDA National Institute of Food and Agriculture. Extension agents provide technical assistance and translate science to farmers, ranchers, and the general public. To find an Extension office near you, visit: http://nifa.usda.gov/partners-and-extension-map?state=All&type=Extension&order=title&sort=asc

International City/County Management Association (ICMA) is an association of professional city and county managers and other employees who serve local governments dedicated to creating and sustaining thriving communities throughout the world. See https://icma.org/

The Johns Hopkins Center for a Livable Future (CLF) engages in research, policy analysis, and technical assistance to advance food system change in local, tribal, institutional, state, federal, and international policy. Operating out of the Bloomberg School of Public Health, it builds the capacity of partner organizations and food policy councils to advocate for a healthy, equitable, resilient, and sustainable food system. See https://clf.jhsph.edu/

National Agricultural Law Center has a library of subject-based reading rooms each with a comprehensive list of electronic resources for an agricultural or food law topic. Links are provided to major statutes, regulations, case law, articles, publications, and other research resources. See https://nationalaglawcenter.org/research-by-topic/

APPENDIX: ORGANIZATIONAL RESOURCES   245

National Association of Conservation Districts (NACD) is a nonprofit organization that represents America's 3,000 conservation districts and the people who serve on their governing boards. NACD promotes responsible natural resources management and conservation by representing locally led conservation districts and their associations through grassroots advocacy, education, and partnerships. See www.nacdnet.org/about-nacd/

National Association of Counties (NACo) serves county elected officials and employees to advocate for county priorities, promote exemplary county policies and practices, nurture leadership skills and expand knowledge networks, optimize county resources, and enrich public understanding of county government. See www.naco.org/what-we-do/about-naco

National Association of Development Organizations (NADO) provides advocacy, education, research, and training for the nation's regional development organizations. The association and its members promote regional strategies, partnerships, and solutions to strengthen the economic competitiveness and quality of life across America's local communities. See www.nado.org/about/

The National Association of Local Boards of Health (NALBOH) informs, guides, and serves as the national voice for local boards of health through education, technical assistance, and advocacy. Its mission is to strengthen and improve public health governance. See www.nalboh.org/page/About

National Association of Towns and Townships (NATaT) serves more than 13,000 towns and townships across the U.S. to enhance the ability of smaller communities to deliver public services, economic vitality, and good government to their citizens. See www.natat.org/

National Association of Regional Councils (NARC) serves as the national voice for regions by advocating for regional cooperation to address community planning and development opportunities and issues. NARC members include regional councils, councils of governments, regional planning and development agencies, Metropolitan Planning Organizations, and other regional organizations. See http://narc.org

North American Food System Network is a professional association of people working together to strengthen local and regional food systems through networking, access to practitioner tools, career guidance, student chapters, webinars, and a curated jobs board. See http://foodsystemsnetwork.org/

The UConn Rudd Center for Food Policy and Health has a policy database that has tracked hundreds of food-related local policies on issues ranging from obesity and diet-related diseases to food access and assistance, farms and gardens, school nutrition menu and package labeling, and food and beverage taxes. See https://healthyfoodpolicyproject.org/policy-database

The Sustainable Development Code is an online source of information on community development that provides resources to help communities develop and amend codes to make them more sustainable. See https://sustainablecitycode.org/about/

United States Conference of Mayors is the official non-partisan organization of cities with populations of 30,000 or more. Members set organizational policies and goals, share information and best practices, and otherwise contribute to the development of urban policy. See www.usmayors.org/

Vermont Law School Center for Agriculture and Food Systems works to address food systems challenges, provides legal services, and develops resources to empower the communities it serves. See www.vermontlaw.edu/academics/centers-and-programs/center-for-agriculture-and-food-systems

BIBLIOGRAPHY

AB-551 Local Government: Urban Agriculture Incentive Zones. Pub. L. No. Assembly Bill No. 551, Chapter 406.

Agha, Mickey, Jeffrey E. Lovich, Joshua R. Ennen, and Brian D. Todd. "Wind, Sun, and Wildlife: Do Wind and Solar Energy Development 'Short-Circuit' Conservation in the Western United States?" *Environmental Research Letters* 15, no. 7 (June 2020): 075004.

Aguilar, Krysten, Lorenzo Alba, Jeff Anderson, Patricia Biever, Jorge Castillo, David Kraenzel, Claudia Mares, and Karim Martinez. "Profiles of Communities of Opportunity: Doña Ana County, New Mexico." *Growing Food Connections*, May 2016.

Alaska Food Policy Council. "Community Food Emergency and Resilience Template." *University of Alaska Fairbanks Cooperative Extension Service*, September 27, 2015.

Albemarle County Code: Chapter 18. Zoning, Section 5: Supplementary Regulations, 2019.

Alkon, Alison Hope, and Teresa Marie Mares. "Food Sovereignty in US Food Movements: Radical Visions and Neoliberal Constraints." *Agriculture and Human Values* 29, no. 3 (September 1, 2012): 347–59.

Allison, Taber D., Jay E. Diffendorfer, Erin F. Baerwald, Julie A. Beston, David Drake, Amanda M. Hale, Cris D. Hein, et al. "Impacts to Wildlife of Wind Energy Siting and Operation in the United States." *Issues in Ecology, Report No. 21.* Washington, DC: Ecological Society of America, Fall 2019.

Alshawaf, Mohammad, Ellen Douglas, and Karen Ricciardi. "Estimating Nitrogen Load Resulting from Biofuel Mandates." *International Journal of Environmental Research and Public Health* 13, no. 5 (May 9, 2016): 478.

American Farmland Trust. "Erie County Agricultural and Farmland Protection Plan." *Farmland Information Center*, October 24, 2012.

———. *2022 Land Trust Survey.* Northampton, MA: American Farmland Trust, forthcoming.

———. *About Planning for Agriculture.* Farmland Information Center.

———. "Farms Under Threat."

———. "Protected Agricultural Lands Database."

American Farmland Trust and Conservation Biology Institute. "San Joaquin Land and Water Strategy: Exploring the Intersection of Agricultural Land & Water Resources in California's San Joaquin Valley." *American Farmland Trust and Conservation Biology Institute*, July 2018.

American Legal Publishing. "San Francisco Administrative Code: Chapter 59: Healthy Food Retailer Ordinance."

American Planning Association. "Agritourism."

———. *Comprehensive Plan Standards for Sustaining Places*. American Planning Association.

———. "Principles of a Healthy, Sustainable Food System."

A Quick Guide to SNAP Eligibility and Benefits. Washington, DC: Center on Budget and Policy Priorities, March 3, 2023.

"Arcadia Center for Sustainable Food and Agriculture."

Arizona Food Systems Network. *Arizona Statewide Food Action Plan 2022–2024*. Arizona Food Systems Network.

Atherton-Bauer, Shanna. "California Climate Investment Programs and Farmland Protection." December 23, 2022a.

———. "Information from Land Conservation Programs Manager." *California Department of Conservation*, December 23, 2022b.

Bailkey, Martin, and Joe Nasr. "From Brownfields to Greenfields: Producing Food in North American Cities." *Community Food Security News* (Fall 2000): 6.

Bayfield County Wisconsin. "Bayfield County, Wisconsin Ordinances Title 5, Chapter 6: Large-Scale Concentrated Animal Feeding Operations." 2016.

Beale, Calvin. "Salient Features of the Demography of American Agriculture." In *The Demography of Rural Life, Publication #64*, edited by David Brown et al. University Park, PA: Northeast Regional Center for Rural Development, 1993.

Beaulieu, Lionel J. "Mapping the Assets of Your Community: A Key Component for Building Local Capacity." *Southern Rural Development Center*.

Belarmino, Emily, Claire Ryan, Qingbin Wang, Meredith Niles, and Margaret Torness. "Impact of Vermont's Food Waste Ban on Residents and Food Businesses." *College of Agriculture and Life Sciences Faculty Publications*, January 30, 2023.

Bellows, Anne C., and Michael W. Hamm. "U.S.-Based Community Food Security: Influences, Practice, Debate." *Journal for the Study of Food and Society* 6, no. 1 (March 1, 2002): 31–44.

Béné, Christophe. "Resilience of Local Food Systems and Links to Food Security – A Review of Some Important Concepts in the Context of COVID-19 and Other Shocks." *Food Security* 12, no. 4 (August 1, 2020): 805–22.

"Beneficial Uses of Manure and Environmental Protection." U.S. Environmental Protection Agency, National Cattlemen's Beef Association, U.S. Poultry & Egg Association, United Egg Producers, National Pork Producers Council, National Milk Producers Federation, August 2015.

Bernard, Chris. "The Work Continues into 2023 | Hunger Free Oklahoma." January 25, 2023.

Boulder County. "Agricultural Lands on Open Space."

Bower, Kelly M., Roland J. Thorpe, Charles Rohde, and Darrell J. Gaskin. "The Intersection of Neighborhood Racial Segregation, Poverty, and Urbanicity and Its Impact on Food Store Availability in the United States." *Preventive Medicine* 58 (January 2014): 33–39.

Boyanton, Megan. "Pandemic Meat Shortage Spurs Calls to Shift Slaughterhouse Rules." October 19, 2020.
Braun, Joachim von, Kaosar Afsana, Louise Ottilie Fresco, Mohamed Hassan, and Maximo Torero. "Food System Concepts and Definitions for Science and Political Action." *Nature Food* 2, no. 10 (October 2021): 748–50.
Brodt, Sonja, Johan Six, Gail Feenstra, Chuck Ingels, and David Campbell. "Sustainable Agriculture." *Nature Education Knowledge* 3, no. 10 (2011): 1.
Bruder, Steven M., and New Jersey State Agriculture Development Committee. "Data on the TDR Program." December 22, 2022.
Bult, Laura. "How 4 Companies Control the Beef Industry." *Vox*, September 29, 2021.
Burlington County Agriculture Development and Board. *Burlington County Comprehensive Farmland Preservation Plan (2022 Update)*. Burlington County, NJ: Burlington County Agriculture Development Board, 2022.
Burlington School Food Project. "Healthy Meals for Better Learning!"
Burns, Christopher, Nigel Key, Sarah Tulman, Allison Borchers, and Jeremy G. Weber. *Farmland Values, Land Ownership, and Returns to Farmland, 2000–2016*. Economic Research Report. Washington, DC: USDA Economic Research Service, February 2018.
Butte County, CA Code of Ordinances: Division 7.—Agricultural Buffers, Chapter 24, Article III, Division 7 §.
Byington, Lillianna. "By the Numbers: Examining the Cost of the Pandemic on the Meat Industry." November 19, 2020.
California Air Resources Board. "2022 Scoping Plan for Achieving Carbon Neutrality." *California Air Resources Board*, November 16, 2022.
California Climate and Action Network. "Sustainable Agricultural Lands Conservation Program (SALCP)."
California Department of Food and Agriculture. "CDFA—OEFI—Conservation Agriculture Planning Grants Program."
California Strategic Growth Council, "Affordable Housing and Sustainable Communities."
CalRecycle. "California's Short-Lived Climate Pollutant Reduction Strategy."
Center for Climate and Energy Solutions. "U.S. State Climate Action Plans." December 2022.
ChangeLab Solutions. "Licensing for Lettuce: A Guide to the Model Licensing Ordinance for Healthy Food Retailers." *ChangeLab Solutions*, 2013.
Chapter 17.20 AGRICULTURE ZONE, Eagle Mountain Municipal Code Ordinance O-33-202 §, 2023.
Chapter 336 – Urban Garden District, Ord. No. 208-07 Cleveland Ohio Municipal Code Chapter 336 §, 2007.
Chapter 69: Agriculture and Markets, Article 25-AA: Agricultural Districts, Consolidated Laws of New York §, 2017.
Chief Executive Office County of Los Angeles. *Fiscal Year 2022–23 Supplemental Budget at a Glance*. Los Angeles, CA: County of Los Angeles.
City and County of San Francisco Department of Public Health. "Application Packet for an Urban Agriculture Incentive Zone." February 20, 2015.
City Council of Minneapolis. Amending Title 10, Chapter 203 of the Minneapolis Code of Ordinances relating to Food Code: Grocery Stores.
City of Baltimore Department of Planning. "Food Policy." March 29, 2016.

City of Fresno Planning and Development. *Fresno General Plan: Chapter 10 Healthy Communities.* Fresno, CA: City of Fresno, December 2014.

City of Fresno the Development and Resource Management Department. Fresno Municipal Code Chapter 15: Citywide Development Code, 2016.

"City of Kansas City, Missouri Parks and Recreation Department Mobile Vending Policy."

City of Minneapolis. "Healthy Living."

———. *Staple Foods Ordinance Fact Sheet: Minneapolis Code of Ordinances. Title 10. Chapter 203. Grocery Stores.* Minneapolis, MN: Minneapolis Department of Health.

City of Portland, OR. "Business Food Scraps Requirement," April 3, 2022.

City of Seattle. Urban Agriculture Ordinance 123378, Ordinance 123378 §, 2010.

"City of Seattle Sweetened Beverage Tax," 2019.

City of Somerville. "City of Somerville Urban Agriculture Zoning Amendment." August 16, 2012.

Claeys, Priscilla. *From Food Sovereignty to Peasants' Rights.* New Haven, CT: Transnational Institute, 2013.

———. "Food Sovereignty and the Recognition of New Rights for Peasants at the UN: A Critical Overview of La Via Campesina's Rights Claims over the Last 20 Years." *Globalizations* 12, no. 4 (July 4, 2015): 452–65.

Clark County Washington. *Clark County Code: 40.240.130 Agricultural Buffer Zones in the General Management Area* (Title 40). Clark County, WA: Unified Development Code, October 26, 2022.

Clark, Jill K., Brian Conley, and Samina Raja. "Essential, Fragile, and Invisible Community Food Infrastructure: The Role of Urban Governments in the United States." *Food Policy* 103 (August 1, 2021): 102014.

Clark, Jill K., Julia Freedgood, Aiden Irish, Kimberley Hodgson, and Samina Raja. "Fail to Include, Plan to Exclude: Reflections on Local Governments' Readiness for Building Equitable Community Food Systems." *Built Environment* 43, no. 3 (September 1, 2017): 315–27.

Clean Water Indiana. "Clean Water Indiana: 2023 Competitive Grants." October 25, 2022.

Cole, Emily J. *Regenerative Agriculture for New England: Sustaining Farmland Productivity in a Changing Climate.* Northampton, MA and Boston, MA: American Farmland Trust and Conservation Law Foundation, August 8, 2022.

Coleman-Jensen, Alisha, Matthew P. Rabbitt, Christian A. Gregory, and Anita Singh. *Household Food Security in the United States in 2021* (Economic Research Report). Washington, DC: USDA Economic Research Service, 2021.

Colorado Department of Agriculture. "Soil Health."

Committee on Budget and Fiscal Review. Senate Bill 862, Pub. L. No. 862, 2014.

COPE: Community Outreach and Patient Empowerment. "COPE Program | Food Access."

Copernicus Climate Change Service. "Global Climate Highlights 2022." *Copernicus*, 2022.

Council of Development Finance Agencies. "CDFA – CDFA Spotlight: Aggie Bonds."

Cromartie, John, and Shawn Bucholtz. "USDA ERS—Defining the 'Rural' in Rural America." *Amber Waves*, June 1, 2008.

Daniels, Thomas L. "Assessing the Performance of Farmland Preservation in America's Farmland Preservation Heartland: A Policy Review." *Society & Natural Resources* 33, no. 6 (June 2, 2020): 758–68.

Davis, Alison, and Simona Balazs. "The Influence of the Agricultural Cluster on the Fayette County Economy." *Community and Economic Development Initiative of Kentucky College of Agriculture, Food, and Environment University of Kentucky*, May 2017.
D.C. Hunger Solutions. "D.C. Healthy Schools Act Brochure." August 1, 2010.
"Delaware: A Small State That Is Big in Agriculture."
Delaware Department of Agriculture—State of Delaware. "Aglands Preservation Program."
DelReal, Jose A., and Scott Clement. "Rural Divide: New Poll of Rural Americans Shows Deep Cultural Divide with Urban Residents." *The Washington Post*, June 17, 2017.
Dempsey, Jennifer, and Laura Barley. "Status of State Purchase of Agricultural Conservation Easement Programs." *American Farmland Trust: Farmland Information Center*, October 25, 2022.
Denver Department of Public Health and Environment. *Denver Food Action Plan*. Denver, CO: Denver Department of Public Health and Environment, June 2018.
Department of Commerce Advisory Committee on Zoning. A Standard State Zoning Enabling Act Under Which Municipalities May Adopt Zoning Regulations, 1928a.
———. "Standard State Zoning Enabling Act and Standard City Planning Enabling Act." *American Planning Association*, 1928b.
Department of Nutrition, Harvard T.H. Chan School of Public Health. "Healthy Eating Plate." *The Nutrition Source*, September 18, 2012.
Dieter, Cheryl A., Molly A. Maupin, Rodney R. Caldwell, Melissa A. Harris, Tamara I. Ivahnenko, John K. Lovelace, Nancy L. Barber, and Kristin S. Linsey. "Estimated Use of Water in the United States in 2015." In *USGS Numbered Series. Estimated Use of Water in the United States in 2015*. Vol. 1441. Circular. Reston, VA: U.S. Geological Survey, 2018.
Dillemuth, Ann, and Kimberley Hodgson. "Incentivizing the Sale of Healthy and Local Food." *Growing Food Connections*, February 2016.
Dimitri, Carolyn, Lydia Oberholtzer, and Andy Pressman. "Urban Agriculture: Connecting Producers with Consumers." Edited by Fabio Verneau and Professor Christopher J. Griffith. *British Food Journal* 118, no. 3 (January 1, 2016): 603–17.
District of Columbia Department of Energy and Enviornment. "Urban Farm Tax Abatement Program | Doee."
Drewitt, Allan L., and Rowena H. W. Langston. "Assessing the Impacts of Wind Farms on Birds." *Ibis* 148, no. s1 (2006): 29–42.
"Economic Impacts of Supermarkets and Other Healthy Food Retail."
Electricity Markets and Policy Group. "Land-Based Wind Market Report."
El Nassar, Naya. *More Than Half of U.S. Population in 4.6 Percent of Counties*. Washington, DC: U.S. Census Bureau, October 24, 2017.
"Emergencies, Disasters and Ohio's Food System | Center for Community and Working Landscapes."
Entrekin, Nyssa. "Information Provided by Nyssa Entrekin, Associate Director, Healthy Food Retail at the Food Trust." January 2023.
Executive Office, County of Los Angeles. "Board of Supervisors."
"Facts About Los Angeles | Discover Los Angeles."
Fair Food Network. "Double Up Food Bucks: 2021 Annual Impact Report."
Falcone, James A., and Michelle A. Nott. "Estimating the Presence of Paved Surface Parking Lots in the Conterminous U.S. from Land Use Coefficients for 1974, 1982, 1992, 2002, and 2012." *U.S. Geological Survey*, 2019.

FAO Agricultural and Development Economics Division. "Food Security (Policy Brief: Issue 2)." *FAO Agricultural and Development Economics Division*, June 2006.

Farmland Information Center. "Farmland Protection Policy Act Fact Sheet." *American Farmland Trust*, July 2022.

———. *Southampton, NY: Lease of Development Rights Enabling Ordinance*. Southampton, NY: Code § 330-248G, 2023.

Farrigan, Tracey. *Rural Poverty & Well-Being*. Washington, DC: USDA Economic Research Service, November 29, 2022.

Federal Food Safety Oversight: Additional Actions Needed to Improve Planning and Collaboration. Washington, DC: U.S. General Accountability Office, December 2014.

Federal Register. "Geographic Preference Option for the Procurement of Unprocessed Agricultural Products in Child Nutrition Programs." April 22, 2011.

Feeding America. "Charitable Food Assistance Participation in 2021." June 2022.

Finegold, Muriel A., Bea Mah Holland, and Tony Lingham. "Appreciative Inquiry and Public Dialogue: An Approach to Community Change." *Public Organization Review* 2, no. 3 (September 1, 2002): 235–52.

Fitzgerald, Kate, Anne Palmer, and Karen Banks. "Understanding the SNAP Program for Food Policy Programs." *Johns Hopkins Center for a Livable Future*, n.d.

Flora, Cornelia Butler, Jan L. Flora, and Susan Fey. *Rural Communities: Legacy and Change*. 2nd ed. Boulder, CO: Westview Press, 2004.

Food Industry Editorial Team. "Who Are the Top 10 Grocers in the United States?" FoodIndustry.Com, June 7, 2022.

"Food Systems—Programs | MU Extension." Michigan State University, Center for Regional Food Systems. https://www.canr.msu.edu/foodsystems/

"Foresight 2.0—Global Panel." May 31, 2019.

Fox, Jessica. "Trading Up: The Ohio River Basin Water Quality Trading Project Is the World's Largest Water Quality Credit Program." *Water Technology*, November 2, 2019.

Francis, Dania V., Darrick Hamilton, Thomas W. Mitchell, Nathan A. Rosenberg, and Bryce Wilson Stucki. "Black Land Loss: 1920–1997." *AEA Papers and Proceedings* 112 (May 2022): 38–42.

Freedgood, Julia, and Jessica Fydenkevez. *Growing Local: A Community Guide to Planning for Agriculture and Food Systems*. Northampton, MA: American Farmland Trust, April 30, 2017.

Freedgood, Julia, Mitch Hunter, Jennifer Dempsey, and Ann Sorenson. *Farms Under Threat: The State of the States*. Northampton, MA: American Farmland Trust, May 13, 2020.

Freedgood, Julia, Marisol Pierce-Quiñonez, and Kenneth Meter. "Emerging Assessment Tools to Inform Food System Planning." *Journal of Agriculture, Food Systems, and Community Development* 2, no. 1 (December 23, 2011): 83–104.

Friedman, Diana. "Clean Energy Farming: Cutting Costs, Improving Efficiencies, Harnessing Renewables. Opportunities in Agriculture." *Sustainable Agriculture Research and Education*, 2012.

Gaither, Cassandra J., Ann Carpenter, Tracy Lloyd McCurty, and Sara Toering. "Heirs' Property and Land Fractionation: Fostering Stable Ownership to Prevent Land Loss and Abandonment." E-General Technical Report SRS-244. Asheville, NC: U.S. Department of Agriculture Forest Service, 2019: 1–105.

Game, Ibrahim, and Richaela Primus. "GSDR 2015 Brief: Urban Agriculture." https://sustainabledevelopment.un.org/content/documents/5764Urban%20Agriculture.pdf

Garfield, Katie, Emma Scott, Kristin Sukys, Sarah Downer, Rachel Landauer, Julianne Orr, Rebecca Friedman, Margaret Dushko, Emily Broad Lieb, and Robert Greenwald. "Mainstreaming Produce Prescriptions: A Policy Strategy Report." *The Center for Health Law and Policy Innovation of Harvard Law School*, 2021.
GISGeography. "13 Open Source Remote Sensing Software Packages." *GIS Geography*, March 15, 2023.
Global Alliance for the Future of Food. "Principles for Food Systems Transformation: A Framework for Action." *Global Alliance for the Future of Food*, June 2021.
Goldschmidt, Walter. *Down on the Farm, New Style*. Yellow Springs, OH: Publisher Not Identified, 1948.
Goldstein, Benjamin, Dimitrios Gounaridis, and Joshua P. Newell. "The Carbon Footprint of Household Energy Use in the United States." *Proceedings of the National Academy of Sciences* 117, no. 32 (August 11, 2020): 19122–30.
Grout, Travis, and Jennifer Ifft. *Approaches to Balancing Solar Expansion and Farmland Preservation: A Comparison across Selected States*. Ithaca, NY: Charles H. Dyson School of Applied Economics and Management, Cornell University, March 2018.
Growing Food Connections. "Cleveland, OH Local Purchasing, Ordinance No. 1660-A-09." *Growing Food Connections Policy Database*.
Haines, Anna L. "What Does Zoning Have to Do with Local Food Systems?" *Journal of Agriculture, Food Systems, and Community Development* 8, no. B (October 17, 2018): 175–90.
Hake, Monica, Emily Engelhard, and Adam Dewey. "Map the Meal Gap 2022: An Analysis of County and Congressional District Food Insecurity and County Food Cost in the United States in 2020." *Feeding America*, 2022.
Hall, Peggy Kirk, and Ellen Essman. *State Legal Approaches to Reducing Water Quality Impacts from the Use of Agricultural Nutrients on Farmland*. Fayetteville, AR: National Agricultural Law Center, May 2019.
Harries, Caroline. "Data on the Pennsylvania Program Supplied by Caroline Harries of the Food Trust." February 23, 2023.
"Healthy Cleveland Nutrition Guidelines." https://farmlandinfo.org/wp-content/uploads/sites/2/2019/09/healthy_cleveland_nutrition_guidelines.pdf
Healthy Food Healthy People Work Group. "Niagara Falls Local Food Action Plan." 2017.
Healthy Food Policy Project. "Municipal Policies to Support Food Access During Emergencies." February 2, 2021.
Held, Lisa. "Just a Few Companies Control the Meat Industry. Can a New Approach to Monopolies Level the Playing Field?" *Civil Eats*. July 4, 2021. https://civileats.com/2021/07/14/just-a-few-companies-control-the-meat-industry-can-a-new-approach-to-monopolies-level-the-playing-field/
Hendrickson, Mary, William D. Heffernan, Philip H. Howard, and Judith B. Heffernan. "Consolidation in Food Retailing and Dairy." *British Food Journal* 103, no. 10 (January 1, 2001): 715–28.
Hess, Daniel Baldwin, and Jeffrey Rehler. "Parking Reform Could Reenergize Downtowns—Here's What Happened When Buffalo Changed Its Zoning Rules." *The Conversation*, June 10, 2021.
Hicke, Jeffrey A., Lucatello Lucatello, Linda D. Mortsch, Jackie Dawson, Mauricio Domínguez Aguilar, and Carolyn A. F. Enquist. "Chapter 14: North America." In *Climate Change 2022: Impacts, Adaptation and Vulnerability. Contribution of Working Group II to the*

Sixth *Assessment Report of the Intergovernmental Panel on Climate Change*, 1929–2042. Cambridge and New York: Cambridge University Press, 2022.

Hillsborough County City-County Planning Commission. *Food Innovation Districts: A Best Practices Guide to Supporting Locally Produced Agriculture and Food Related Businesses in Hillsborough County*. Tampa, FL: Hillsborough County City-County Planning Commission, n.d.

Hinrichs, C. Clare, and Thomas A. Lyson, eds. *Remaking the North American Food System*. Lincoln: University of Nebraska Press, 2008.

Hoey, Lesli, Lilly Fink Shapiro, Kathryn Colasanti, Alex Judelsohn, Mrithula Shantha Thirumalai Anandanpillai, and Keerthana Vidyasagar. "Participatory State and Regional Food System Plans and Charters in the U.S.: A Summary of Trends and National Directory." *Michigan State University's Center for Regional Food Systems*, August 2021.

House Bill 223: Property Tax Credit—Urban Agricultural Property—Applicability, Pub. L. No. 223.

Hunter, Mitch, Ann Sorenson, Theresa Nogeire-McRae, Scott Beck, Stacy Shutts, and Ryan Murphy. *Farms Under Threat 2040: Choosing an Abundant Future*. Washington, DC: American Farmland Trust, June 29, 2022.

Indian Land Tenure Foundation. "Issues: Land Tenure Issues." https://iltf.org/land-issues/issues/

Information Provided by Nyssa Entrekin, Caroline Haynes, and Julia Koprak in an interview on January 6, 2023, and a Series of Emails between January 3, and March 3, 2023.

Inman, Mason. "Wind Turbines May Help Crops on Farms, Research Says." *National Geographic*, December 21, 2011.

Inwood, Shoshanah, Zoë Plakias, Jill K. Clark, Nicole Wright, Aiden Irish, and Josh D. Vittie. *Preparing for Food System Resiliency in Ohio: Policy and Planning Lessons from COVID-19*. Columbus, OH: College of Food, Agricultural and Environmental Sciences & John Glenn College of Public Affairs, January 2022.

Iowa Department of Natural Resources. "Winter Manure Application: Guide to Frozen and Snow-Covered Ground Rules." May 2021.

Iowa State Legislature. §456A.38 Lease to Beginning Farmers.

Jensen, Eric, Nicholas Jones, Megan Rabe, Beverly Pratt, Lauren Medina, Kimberly Orozco, and Lindsay Spell. *The Chance That Two People Chosen at Random Are of Different Race or Ethnicity Groups Has Increased Since 2010*. Washington, DC: U.S. Census Bureau, August 12, 2021.

Johns Hopkins Center for a Livable Future. "Food Policy Councils Directory." *Johns Hopkins Bloomberg School of Public Health*. https://clf.jhsph.edu/projects/food-policy-networks

Johnson, Renée, and Jim Monke. *Farm Bill Primer: What Is the Farm Bill?* (IF12047, Version 4, Updated). Washington, DC: Congressional Research Service, February 22, 2023.

Karetny, Jane, Casey Hoy, Kareem Usher, Jill Clark, and Maria Conroy. "Planning Toward Sustainable Food Systems: An Exploratory Assessment of Local U.S. Food System Plans." *Journal of Agriculture, Food Systems, and Community Development* 11, no. 4 (September 2, 2022): 115–38.

Katchova, Ani L., and Mary Clare Ahearn. "Dynamics of Farmland Ownership and Leasing: Implications for Young and Beginning Farmers." *Applied Economic Perspectives and Policy* 38, no. 2 (2016): 334–50.

Kelly, Kate, and Kim Delfino. *Smart from the Start: Responsible Renewable Energy Development in the Southern San Joaquin Valley*. Washington, DC: Defenders of Wildlife, 2012.

Ketso. "Workshops & Engagement | Ketso | Ketso Workshops." https://ketso.com/
Kinder, Molly, Laura Stateler, and Julia Du. "Windfall Profits and Deadly Risks." *Brookings Institution*, November 2020.
Kirschenmann, Fred, G. W. Stevenson, Frederick Buttel, Thomas A. Lyson, and Mike Duffy. "Why Worry About the Agriculture of the Middle?" In *Food and the Mid-Level Farm*, edited by Thomas A. Lyson, G. W. Stevenson, and Rick Welsh, 3–20. Cambridge: The MIT Press, 2008.
Kuethe, Todd H., Jennifer Ifft, and Mitch Morehart. "The Influence of Urban Areas on Farmland Values." *Choices Quarter* 2 (2011).
Kunstler, James Howard. *Geography of Nowhere*. New York: Free Press, 1994.
Lakhani, Nina, Aliya Uteuova, and Alvin Chang. "The Illusion of Choice: Five Stats That Expose America's Food Monopoly Crisis." *The Guardian*, July 18, 2021, sec. Environment.
Leeuwis, Cees, Birgit K. Boogaard, and Kwesi Atta-Krah. "How Food Systems Change (or Not): Governance Implications for System Transformation Processes." *Food Security* 13, no. 4 (August 1, 2021): 761–80.
Lennertz, Bill, and Aarin Lutzenhiser. *The Charrette Handbook*. 2nd ed. New York: Routledge, 2017.
Lev, Larry, and Garry Stephenson. "Dot Posters: A Practical Alternative to Written Questionnaires and Oral Interviews." *Journal of Extension, Tools of the Trade* 37, no. 5 (October 1999).
LII/Legal Information Institute. "7 U.S. Code § 3103—Definitions."
Lobao, Linda M. *Locality and Inequality: Farm and Industry Structure and Socioeconomic Condition*. Albany: SUNY Press, 1990.
Los Angeles County Chief Sustainability Office. "Our County—Los Angeles Countywide Sustainability Plan." OurCounty, August 6, 2019.
Lyson, Thomas A., Robert J. Torres, and Rick Welsh. "Scale of Agricultural Production, Civic Engagement, and Community Welfare*." *Social Forces* 80, no. 1 (September 1, 2001): 311–27.
MacDonald, James M., Robert A. Hoppe, and Doris Newton. "Three Decades of Consolidation in U.S. Agriculture." *USDA Economic Research Service*, March 2018.
Mackey, Rachel. *Supplemental Nutrition Assistance Program (SNAP) Reauthorization and Appropriations* (Policy Brief). Washington, DC: National Association of Counties, February 1, 2023.
Magdoff, Fred, and Harold van Es. *Building Soils for Better Crops*. 4th ed. Sustainable Agriculture Network Handbook Series, Book 10. College Park, MD: Sustainable Agriculture Research & Education Program, 2021.
Maillacheruvu, Sara Usha. *The Historical Determinants of Food Insecurity in Native Communities*. Washington, DC: Center on Budget and Policy Priorities, October 4, 2022.
Maine Department of Agriculture, Conservation & Forestry. "Voluntary Municipal Farm Support Program: Farmland Protection Program: M.R.S. 7 Ch2-C."
Mair, Julie Samia, Matthew W. Pierce, and Stephen P. Teret. "The City Planner's Guide to the Obesity Epidemic: Zoning and Fast Food." *The Center for Law and the Public's Health at Johns Hopkins & Georgetown Universities*, October 2005.
———. "The City Planner's Guide to the Obesity Epidemic: Zoning and Fast Food," n.d.
Marshall, Morgan. "Extension Local Food Program Team LFPT Annual Report 2021–2022." *NC State Extension Local Food Program*.

Martinez, Steve, Michael Hand, Michelle Da Pra, Susan Pollack, Katherine Ralston, Travis Smith, Stephen Vogel, et al. *Local Food Systems: Concepts, Impacts, and Issues* (ERR 97). Washington, DC: USDA Economic Research Service, May 2010.

Maryland Agriculture & Resource-Based Industry Development Corporation. "Next Generation Farmland Acquisition Program." https://www.marbidco.org/_pages/programs_land_preservation/rural_land_preservation_programs_nextgen_ngfap.htm

Maryland Department of Agriculture. "Agricultural Nutrient Management Program." *Maryland.gov Enterprise Agency Template.* https://mda.maryland.gov/resource_conservation/Pages/nutrient_management_overview.aspx

———. "Strategic Plan for Maryland Agriculture." *Maryland Department of Agriculture,* December 1, 2019.

Massachusetts Department of Public Health, Bureau of Community Health and Prevention. "Mass in Motion | Mass.Gov." https://www.mass.gov/about-mass-in-motion

Mathias, Chris. Interview with Chris Mathias, Scott County IA Planning Director, December 12, 2022.

Melillo, Jerry M., Gary Wynn Yohe, and Terese Richmond. *Climate Change Impacts in the United States: The Third National Climate Assessment* 10.7930/J0z31wj2OA.Mg." Washington, DC: U.S. Global Change Research Program, 2014.

Mittal, Anuradha. *The 2008 Food Price Crisis: Rethinking Food Security Policies* (Research Papers for the Intergovernmental Group of Twenty-Four on International Monetary Affairs and Development). New York and Geneva: United Nations Conference on Trade and Development, June 2009.

Moorman, Christopher E., Steven M. Grodsky, and Susan P. Rupp. *Renewable Energy and Wildlife Conservation.* Baltimore, MD: Johns Hopkins University Press, 2019.

Multnomah County, Department of County Human Services, SUN Service System. "SUN Food Sites." *Multnomah County,* December 23, 2016.

"Municipal Vulnerability Preparedness (MVP) Program | Mass.Gov." https://www.mass.gov/municipal-vulnerability-preparedness-mvp-program

Myers, Samuel S., Antonella Zanobetti, Itai Kloog, Peter Huybers, Andrew D. B. Leakey, Arnold J. Bloom, Eli Carlisle, et al. "Increasing CO2 Threatens Human Nutrition." *Nature* 510, no. 7503 (June 2014): 139–42.

National Agricultural Law Center. "Water Law Overview." https://nationalaglawcenter.org/overview/water-law/

National Agricultural Law Center Research Staff. *Agritourism – National Agricultural Law Center: States' Agritourism Statutes.* Fayetteville, AR: National Agricultural Law Center, February 9, 2021.

National Center for Chronic Disease Prevention and Health Promotion (NCCDPHP). "About Chronic Diseases." CDC, July 21, 2022.

National Conference of State Legislatures. "Number of Legislators and Length of Terms in Years." April 19, 2021.

National Farm to School Network. "State Farm to School Policy Handbook 2002–2020." *National Farm to School Network, Vermont Law School's Center for Agriculture and Food Systems,* July 2021.

National Farmers Market Directory. "Find a Local Farmers Market Near You." and another: https://nfmd.org/browse/

National Policy & Legal Analysis Network to Prevent Childhood Obesity. "Healthy Mobile Vending Policies: A Win-Win for Vendors and Childhood Obesity Prevention Advocates." *National Policy & Legal Analysis Network to Prevent Childhood Obesity*, 2013.

Navajo Epidemiology Center. "Understanding the Healthy Diné Nation Act of 2014."

Nelson, Gerald C., Mark W. Rosegrant, Jawoo Koo, and Robertson. *Climate Change: Impact on Agriculture and Costs of Adaptation*. Washington, DC: International Food Policy Research Institute, 2009.

Nerds for Earth. "State Healthy Soil Policy Map." *Nerds for Earth*, August 27, 2019.

New Mexico Department of Agriculture. "Healthy Soil Program." and another: https://nmdeptag.nmsu.edu/healthy-soil-program.html

New Orleans Fresh Food Retailer Initiative: Information Sheet. New Orleans, LA: City of New Orleans, n.d.

New York City Department of City Planning. "Zoning Districts & Tools: Rules for Special Areas: FRESH Food Stores-DCP."

New York City Mayor's Office of Contract Services. "New York State Food Purchasing Guidelines, 2012 (Revised April 7, 2015)."

New York State Department of Agriculture and Markets. "Climate Resilient Farming." https://agriculture.ny.gov/soil-and-water/climate-resilient-farming

———. "Farmland Protection Planning Grants Program." https://agriculture.ny.gov/land-and-water/farmland-protection-planning-grants-program

Nixon, Laura, Pamela Mejia, Lori Dorfman, Andrew Cheyne, Sandra Young, Lissy C. Friedman, Mark A. Gottlieb, and Heather Wooten. "Fast-Food Fights: News Coverage of Local Efforts to Improve Food Environments Through Land-Use Regulations, 2000–2013." *American Journal of Public Health* 105, no. 3 (March 2015): 490–96.

North Carolina Department of Agriculture & Consumer Services. "Soil & Water Conservation Division—Cost Share Programs: Agricultural Cost Share Program (ACSP)."

North Iowa Area Council of Governments. Cerro Gordo County Ordinance Number 15: Zoning Ordinance of Cerro Gordo County, Iowa: Revised through December 13, 2022.

Northampton County, NC Voluntary Agricultural District Ordinance.

Oregon Department of Land Conservation and Development. "A Summary of Oregon's Statewide Planning Goals."

———. "Oregon Statewide Planning Goals and Guidelines." *Oregon Department of Land Conservation and Development*, July 2019.

Oregon Legislative Assembly. Chapter 215—County Planning; Zoning; Housing Codes, 2021 Edition.

ORS 215.780—Minimum lot or parcel sizes, Pub. L. No. Oregon Laws, Oregon Administrative Rules, Oregon Revised Statutes, Volume 6, Title 20, Chapter 215, Section 215.780.

"Partition of Heirs Property Act—Uniform Law Commission." https://www.uniformlaws.org/committees/community-home?CommunityKey=50724584-e808-4255-bc5d-8ea4e588371d

Paustian, Keith, Johannes Lehmann, Stephen Ogle, David Reay, G. Philip Robertson, and Pete Smith. "Climate-Smart Soils." *Nature* 532, no. 7597 (April 2016): 49–57.

Pearce, Dustin, James Strittholt, Terry Watt, and Ethan Elkind. *A Path Forward: Identifying Least-Conflict Solar PV Development in California's San Joaquin Valley.* Berkeley, CA: University of California, Berkeley Law, May 1, 2016.

Pennsylvania General Assembly. Act of July 25, 2007, P.L. 373, No. 55 Cl. 72—Tax Reform Code of 1971—Omnibus Amendments, 2007.

Pennsylvania State Conservation Commission. "PA REAP FY 2019 Annual Report."

Pippidis, Maria, Bonnie Braun, Jesse M. Ketterman, Shoshanah Inwood, and Nicole Wright. *Engaging Communities Through Issues Forums: A How-to Guide for Onsite and Online Community Engagement.* Kansas City, MO: Extension Foundation, 2022.

Porter, Michael E. "New Strategies for Inner-City Economic Development." *Economic Development Quarterly* 11, no. 1 (February 1, 1997): 11–27.

Pörtner, Hans-O., Debra C. Roberts, Helen Adams, Elvira Poloczanska, and Katja Mintenbeck, eds. "Summary for Policy Makers." In *Climate Change 2022: Impacts, Adaptation and Vulnerability. Contribution of Working Group II to the Sixth Assessment Report of the Intergovernmental Panel on Climate Change,* 3–33. Cambridge and New York: Cambridge University Press, 2022.

Pothukuchi, Kameshwari, and Jerome L. Kaufman. "Placing the Food System on the Urban Agenda: The Role of Municipal Institutions in Food Systems Planning." *Agriculture and Human Values* 16, no. 2 (June 1, 1999): 213–24.

———. "The Food System: A Stranger to the Planning Field." *Journal of the American Planning Association* 66, no. 2 (June 30, 2000): 113–24.

"Prairie Crossing | A Conservation Community in Grayslake, Illinois." https://prairie-crossing.com/

"Produce Prescription (Rx) Program – Fresh Bucks Indy." https://freshbucksindy.org/produce-rx/#:~:text=Produce%20Rx%20is%20a%20partnership,diseases%20while%20improving%20nutrition%20security.

Rangarajan, Anu, and Molly Riordan. "The Promise of Urban Agriculture, National Study of Commercial Farming in Urban Areas." *USDA Agricultural Marketing Service and Cornell University Small Farms Program,* August 2019.

Rao, Mayuree, Ashkan Afshin, Gitanjali Singh, and Dariush Mozaffarian. "Do Healthier Foods and Diet Patterns Cost More Than Less Healthy Options? A Systematic Review and Meta-Analysis." *BMJ Open* 3, no. 12 (December 1, 2013): e004277.

Regulation and Funding: Natural Resources Department, §456A.38: Lease to Beginning Farmers Program.

"Residents: County of Los Angeles." https://lacounty.gov/residents/

Rhode Island Office of Energy Resources and The Division of Statewide Planning. "Rhode Island Renewable Energy Guidelines: Solar Energy Systems (SES) Model Ordinance Templates, Zoning and Taxation." February 2019.

RS 3:304 Master Farmer Certification, Louisiana Laws – Louisiana State Legislature §, 2008.

Ruhf, Kathryn Z., and Kate Clancy. "A Regional Imperative: The Case for Regional Food Systems." *Thomas A. Lyson Center for Civic Agriculture and Food Systems,* September 20, 2022.

Saitone, Tina L., and Patrick W. McLaughlin. "Women, Infants and Children (WIC) Program Redemptions at California Farmers' Markets: Making the Program Work for Farmers and Participants." *Renewable Agriculture and Food Systems* 33, no. 4 (August 2018): 334–46.

Sandt, Anders Van, Sarah A. Low, and Dawn Thilmany. "Exploring Regional Patterns of Agritourism in the U.S.: What's Driving Clusters of Enterprises?" *Agricultural and Resource Economics Review* 47, no. 3 (December 2018): 592–609.
Santo, Raychel, Caitlin Misiaszek, Karen Bassarab, Darriel Harris, and Anne Palmer. "Pivoting Policy, Programs, and Partnerships: Food Policy Councils Response to the Crises of 2020." *Johns Hopkins Center for a Livable Future*, 2021.
"SB717 – Authorizes a Tax Credit for Urban Farms." 2022. https://www.senate.mo.gov/22info/BTS_Web/Bill.aspx?SessionType=R&BillID=71259647#:~:text=Current%20Bill%20Summary&text=The%20tax%20credit%20shall%20not,%24100%2C000%20in%20any%20calendar%20year. August 8, 2022
Schilling, Brian J., J. Dixon Esseks, Joshua M. Duke, Paul D. Gottlieb, and Lori Lynch. "The Future of Preserved Farmland: Ownership Succession in Three Mid-Atlantic States." *Journal of Agriculture, Food Systems, and Community Development* 5, no. 2 (February 24, 2015): 129–53.
Scott County, Iowa. "2007 Scott County Comprehensive Plan." July 17, 2015.
———. "Scott County Iowa Economy & Market," June 7, 2016a.
———. Zoning Ordinance for Unincorporated Scott County, 2016b.
Seattle Department of Neighborhoods. "Food Equity Fund – Neighborhoods | Seattle.Gov."
Sec. 8 – 2.404. Agricultural Conservation and Mitigation Program., Title 8, Chapter 2, Section 8 – 2.404 Yolo County Code of Ordinances §.
Select Subcommittee on the Coronavirus Crisis Majority Staff. "Coronavirus Infections and Deaths Among Meatpacking Workers at Top Five Companies Were Nearly Three Times Higher than Previous Estimates." *Congress of the United States, House of Representatives*, October 27, 2021.
Serra, Cara. *Regional Resiliency Action Plan*. Pinellas Park, FL: Tampa Bay Regional Planning Council, November 2022.
Sideroff, Desiree. "Getting to Grocery: Tools for Attracting Healthy Food Retail to Underserved Neighborhoods." *ChangeLab Solutions*, 2012.
Sonoma County, California. Chapter 30 – Agriculture. Article II. – Right to Farm, Ord. No. 5203 § 5, 1999 §.
Sorenson, Ann, Julia Freedgood, Jennifer Dempsey, and David Theobald. *Farms Under Threat: The State of America's Farmland*. Washington, DC: American Farmland Trust, 2018.
State of California Governor's Office of Planning and Research, and California Strategic Growth Council. "Grant Funding Opportunity: Tribal Government Challenge Planning Grant Program." March 13, 2020.
Statista. "Number of U.S. Cities, Towns, Villages by Population Size 2019."
Suneson, Grant. "What Are the 25 Lowest Paying Jobs in the US? Women Usually Hold Them," *USA TODAY*, accessed February 2, 2023, https://www.usatoday.com/story/money/2019/04/04/25-lowest-paying-jobs-in-us-2019-includes-cooking-cleaning/39264277/.
Takle, Gene, and Ed Adcock. "Iowa State University Research Finds Wind Farms Positively Impact Crops." *Iowa State University College of Agriculture and Life Sciences*, March 5, 2018.
Tendall, Danielle M., J. Joerin, B. Kopainsky, P. Edwards, A. Shreck, Q. B. Le, P. Kruetli, M. Grant, and J. Six. "Food System Resilience: Defining the Concept." *Global Food Security* 6 (October 1, 2015): 17–23.

Terry, Georgena. "State Pollinator-Friendly Solar Initiatives." *Clean Energy States Alliance*, January 2020.

Texas Commission on Environmental Quality. Subchapter B: Concentrated Animal Feeding Operations §§321.31–321.47, Chapter 321 – Control of Certain Activities by Rule §, 2014.

The Food Trust. *Healthier Corner Stores: Positive Impacts and Profitable Changes*. Philadelphia, PA: The Food Trust, 2014.

The New York City Administrative Code: § 17–307 Licenses, Permits Required; Restrictions; Term.

"The Sarasota County Comprehensive Plan: A Planning Tool for the Future of Sarasota County. Volume 1: Goals, Objectives, & Policies." *Sarasota County Board of County Commissioners*, October 25, 2016.

Thilmany, Dawn, Elizabeth Canales, Sarah A. Low, and Kathryn Boys. "Local Food Supply Chain Dynamics and Resilience during COVID-19." *Applied Economic Perspectives and Policy* 43, no. 1 (2021): 86–104.

Town of Lowville, NY: Anaerobic Digesters, 2013.

Township of Chesterfield, Burlington County, New Jersey. "Township of Chesterfield, NJ: Article XVII Voluntary TDR Program Procedural Requirements." *Township of Chesterfield, NJ Code*.

Troy, Mike. "Pandemic-Fueled Record Growth in 2020: The PG 100." *Progressive Grocer*, May 17, 2021.

"UConn Rudd Center for Food Policy and Health." https://uconnruddcenter.org/

United Nations Department of Economic and Social Affairs, Population Division. "World Population Prospects 2022: Summary of Results." *United Nations Department of Economic and Social Affairs, Population Division*, June 11, 2022.

United States Environmental Protection Agency. "Agriculture." *Inventory of U.S. Greenhouse Gas Emissions and Sinks: 1990–2020* 5–60 (2022).

Urban Land Institute, ed. *America in 2015: A ULI Survey of Views on Housing, Transportation, and Community*. Washington, DC: Urban Land Institute, 2015.

U.S. Census Bureau. "1920 Census: Volume 5. Agriculture, Reports for States, Chapter 5, Farm Statistics by Race, Nativity, and Sex of Farmer." *U.S. Census Bureau*, 1920.

———. "Historical Population Change Data (1910–2020)." *Census.gov*, April 26, 2021.

———. "Metropolitan and Micropolitan Statistical Areas Population Totals and Components of Change: 2020–2021." *Census.gov*.

———. "SNAP Benefits Recipients in Philadelphia County/City, PA." *FRED, Federal Reserve Bank of St. Louis*. FRED, Federal Reserve Bank of St. Louis.

———. "State Population Totals and Components of Change: 2020–2022." *Census.gov*.

———. "U.S. Census Bureau QuickFacts: United States."

———, and America Counts Staff. "What Is Rural America? One in Five Americans Live in Rural Areas." *Census.gov*, August 9, 2017.

U.S. Composting Council. "Model Compost Rules Template (MCRT) Version 2.0." https://www.compostingcouncil.org/page/ModelRuleTemplate

U.S. Department of Agriculture. "Food Waste FAQs." https://www.usda.gov/foodwaste/faqs

———, and President's Council on Environmental Quality. *National Agricultural Lands Study*. Washington, DC: National Agricultural Lands Study, January 1, 1981.

U.S. Department of Energy, Office of Energy Efficiency and Renewable Energy. "Solar Futures Study." 2021.
U.S. Department of Interior. "Land Buy-Back Program for Tribal Nations: Fractionation." February 25, 2019.
U.S. Department of Interior, Bureau of Land Management. "Livestock Grazing on Public Lands."
U.S. Environmental Protection Agency. "Community Solar: An Opportunity to Enhance Sustainable Development on Landfills and Other Contaminated Sites." December 2016.
———. "Local Foods, Local Places." *Overviews and Factsheets*, June 6, 2014.
———. "Reducing the Impact of Wasted Food by Feeding the Soil and Composting." *Overviews and Factsheets*, August 12, 2015.
USDA. "Urban Agriculture."
USDA Agricultural Marketing Services. "Packers and Stockyards Division Annual Report 2019." *USDA Agricultural Marketing Service*, 2019.
USDA Economic Research Service. "Food Security Status of U.S. Households in 2021." *USDA ERS Key Statistics & Graphics*.
———. "Irrigation & Water Use." May 6, 2022a.
———. "USDA ERS – Food Security and Nutrition Assistance." October 18, 2022b.
———. "USDA ERS – Food Prices and Spending." January 6, 2023.
———. "USDA ERS – Farm Bill Spending."
———. "USDA ERS – Farming and Farm Income."
———. "USDA ERS – Feed Grains Sector at a Glance."
———. "USDA ERS – Food Security in the U.S.: Key Statistics & Graphics." https://www.ers.usda.gov/topics/food-nutrition-assistance/food-security-in-the-u-s/key-statistics-graphics/ Last updated October 25, 2023
USDA Food and Nutrition Service. "Barriers That Constrain the Adequacy of Supplemental Nutrition Assistance Program (SNAP) Allotments (Summary)." *USDA Food and Nutrition Service*, June 2021.
———. "Farm to School Census." March 19, 2022.
USDA Food and Nutrition Service's Supplemental Nutrition Assistance Program. "Healthy Corner Stores: Making Corner Stores Healthier Places to Shop." *USDA Food and Nutrition Service's Supplemental Nutrition Assistance Program*, 2016.
———. "Farm to School Census and Comprehensive Review." July 15, 2021.
———. "Farmers' Markets Accepting SNAP Benefits | Food and Nutrition Service." January 5, 2023.
USDA National Agricultural Statistics Service. *2012 Census of Agriculture Highlights: Direct Farm Sales of Food.* Washington, DC: USDA National Agricultural Statistics Service, December 2016.
———. "2017 Census of Agriculture County Profile: Scott County Iowa." *USDA National Agricultural Statistics Service*, 2017a.
———. "2017 Census of Agriculture, County Profile: Wayne County, Ohio." 2017b.
———. "2017 U.S. Census of Agriculture, Introduction." *USDA National Agricultural Statistics Service*, 2017c.
———. "Direct Farm Sales of Food: Results from the 2020 Local Food Marketing Practices Survey." April 2022.
———. "2022 Census of Agriculture Highlights: 'Farm Economics'." *USDA National Agricultural Statistics Service*, 2024.

USDA, National Agricultural Statistics Service, 2022 *Census of Agriculture – National Data. Table 52: Selected Producer Characteristics* (Washington, DC: USDA-NASS, February 2024).

USDA National Organic Program Agricultural Marketing Service. "Introduction to Organic Practices, The NOP Organic Insider." September 2015.

USDA Natural Resources Conservation Service. "Conservation Practice Standard Nutrient Management (Code 590)." May 2019.

———. "Conservation Practices on Cultivated Cropland: A Comparison of CEAP I and CEAP II Survey Data and Modeling – Summary of Findings." *USDA Natural Resources Conservation Service*, January 2022.

———. "M_310_523_C_April 2013 – Subpart C – Important Farmland Soils."

Valliant, Julia, and Julia Freedgood. "Land Access Policy Incentives: Emerging Approaches to Transitioning Farmland to a New Generation." *Journal of Agriculture, Food Systems, and Community Development* 9, no. 3 (May 15, 2020): 71–78.

Valliant, Julia C. D., Kathryn Z. Ruhf, Kevin D. Gibson, J. R. Brooks, and James R. Farmer. "Fostering Farm Transfers from Farm Owners to Unrelated, New Farmers: A Qualitative Assessment of Farm Link Services." *Land Use Policy* 86 (July 1, 2019): 438–47.

Vermont Department of Environmental Conservation. "Clean Water Initiative."

Vermont General Assembly. Title 6: Agriculture, Chapter 215: Agricultural Water Quality, Subchapter 2: Water Quality; Required Agricultural Practices and Best Management Practice including §§ 4810, 4810a, and 4811.

Vermont Required Agricultural Practices Rule for the Agricultural Nonpoint Source Pollution Control Program (Act 64 of the Vermont General Assembly, 2015 Session). Montpelier, VT: Vermont Agency of Agriculture, Food & Markets, November 23, 2018.

Vermont Sustainable Jobs Fund (VSJF). "Vermont Farm to Plate Strategic Plan."

"Virginia Administrative Code – Title 9. Environment – Agency 25. State Water Control Board – Chapter 32. Virginia Pollution Abatement (VPA) Permit Regulation."

Virginia Department of Agriculture and Consumer Services. "AFID Planning Grants."

Walston, Leroy J., Shruti K. Mishra, Heidi M. Hartmann, Ihor Hlohowskyj, James McCall, and Jordan Macknick. "Examining the Potential for Agricultural Benefits from Pollinator Habitat at Solar Facilities in the United States." *Environmental Science & Technology* 52, no. 13 (July 3, 2018): 7566–76.

Walthal, Charles L., Jerry Hatfield, P. Backlund, Laura Lengnick, E. Marshall, M. Walsh, S. Adkins, et al. *Climate Change and Agriculture in the United States: Effects and Adaptation* (USDA Technical Bulletin 1935). Washington, DC: USDA ARS, 2012.

Webber, David, Alan Balsam, and Bonita Oehlke. "The Massachusetts Farmers' Market Coupon Program for Low Income Elders." *American Journal of Health Promotion* 9, no. 4 (March 1, 1995): 251–53.

Weber, Christopher L., and H. Scott Matthews. "Food-Miles and the Relative Climate Impacts of Food Choices in the United States." *Environmental Science & Technology* 42, no. 10 (May 15, 2008): 3508–13.

Whitt, Christine, Noah Miller, and Ryan Olver. "America's Farms and Ranches at a Glance: 2022 Edition." *USDA Economic Research Service*, December 2022.

"Williamson Act Program Overview."

"Wind Vision Detailed Roadmap Actions: 2017 Update." *U.S. Department of Energy, Office of Energy Efficiency and Renewable Energy*, May 2018.

Wisconsin Department of Natural Resources. "Composting Rules and Regulations in Wisconsin."
Wood, Dustin A., Amy G. Vandergast, Kelly R. Barr, Rich D. Inman, Todd C. Esque, Kenneth E. Nussear, and Robert N. Fisher. "Comparative Phylogeography Reveals Deep Lineages and Regional Evolutionary Hotspots in the Mojave and Sonoran Deserts." *Diversity and Distributions* 19, no. 7 (2013): 722–37.
World Population Review. "Arable Land by Country 2023."
Zeuli, Kim, Austin Nijhuis, and Pete Murphy. *Resilient Food Systems, Resilient Cities: Recommendations for the City of Boston.* Boston, MA: Initiative for a Competitive Inner City, May 2015.
Zoning Ordinance of Lexington-Fayette County, Kentucky, Ordinance No. 122-2022, Article 1 – General Provisions and Definitions, 2022.

INDEX

Note: Page numbers in *italics* indicate a figure and page numbers in **bold** indicate a table on the corresponding page. Page numbers followed by "n" with numbers refer to notes.

Affordable Housing and Sustainable Communities (AHSC) Program 168
Agricultural Conservation Easement Program (ACEP) 114, 116, 163; Agricultural Land Easements (ALE) 114, 164
agricultural districts (agricultural security areas) 160
Agricultural Improvement Act of 2018 108–109, *108*
Agricultural Overlay Zones 131, 137
Agricultural Protection Zones (APZ) 133–136
agricultural viability support and conservation policies/programs 181–182; aggie bond programs 184; agricultural advisory boards and commissions 183–184; agricultural economic development authorities 184–185; "buy local" branding campaigns 186–187; conservation promotion 167; cooperative extension 185–186, *185*; farmers' markets 187; farm-to-school programs 187–188, *188*; farm viability programs 186; food hubs 187; Hawaii's Agricultural Parks Program 223; market support 186–188, *188*; new unit notification 182; right-to-farm laws and ordinances 182; and solar energy siting 226–228
Agrihoods (Development Supported Agriculture) 125
agritourism (agrotourism/agritainment)139–140
AHSC *see* Affordable Housing and Sustainable Communities
Alaska Food Policy Council 67, 75n19

INDEX 265

ALE *see* Agricultural Land Easements
American Farmland Trust (AFT) 23n38, 45, 75n12, 98, 114, 222, 227, 232, 243
American Farmland Trust's Farmland Information Center/Farmland Information Center (FIC) 171–172, 243; Agricultural District Provisions by Outcome *161*; Direct Sales of Food Produced om Urban-Influenced Countries *15*; Federal Impact on Agricultural Land Conversion *107*; Food Produced in Urban-Influenced Counties *73*
American Planning Association (APA) 9, 62, 78, 243
American Public Health Association 78, 213
animal cruelty 150
animal feeding operations (AFOs) 118, 176; *see also* concentrated animal feeding operations (CAFOs)
animal welfare 29, 149–151
appreciative inquiry 93
APZ *see* Agricultural Protection Zones

Baltimore Food Policy Initiative (BFPI) 213–214
biodiversity 6, 17, 36, 78, 117
Brewington vs. Glickman 31
Bureau of Land Management (BLM) 106, 117, 222

CAFOs *see* concentrated animal feeding operations
California's San Joaquin Valley 72, 230, 231
cannabis, commercial 20, 217; cannabis plant 233–237, *234*, **235**; canopy 236; community guidance 230–231; energy and light 236

Carson, Rachel, *Silent Spring* 5, 35
CDFI *see* Community Development Financial Institution
CFPs *see* Community Food Projects
ChangeLab Solutions 216n27, 244
Chesapeake Bay Watershed water quality 177–178, *177*
Child Nutrition Act *111*, 112
CIG *see* Conservation Innovation Grants
CivicWell 244
Clean Water Act (CWA) 11, 149, 173–176
Clean Water Indiana (CWI) 173, 189n23
climate action plans (CAPs) 65
climate change 9, 12, 26, 37, 43, 47, 65, 172, 238
cluster development (Conservation/Open Space Development) 126
Colorado Department of Agriculture 171, 189n17
Commodity Supplemental Food Program 110
community/ies: capitals 79–80; cohesive 7; cultural diversity of 53; engagement 238; gardens 18, 38, 72, 131, 152, 194, 221–224; guidance 230; Indigenous 4; rural *see* rural community; tribal 72; underserved 196–197, 199–200; urban 70–71, 74, 81, 199
Community Capitals Framework 74, 79
Community Development Financial Institution (CDFI) 199, 226
Community Food Projects (CFPs) 110–111
community food security 17, 20, 53, 81, 145, 151; coordinating emergency food 212–214; emergency food systems 211–214, *211*; food marketing and

266   INDEX

sales 208; food policy councils 194–195; funding for emergency food 212; healthy corner store initiatives 196–199, *196*; healthy food financing initiatives (HFFI) 199–200; healthy retail licensing 200–201; healthy retail programs 195–205; mobile food vending allowances 201–202; nutrition education and promotion 205–211; nutrition guidelines 206–208, *207*; nutrition incentive programs 202–205; school wellness policy 209–211, *209*
community land trusts 18
Community Supported Agriculture (CSA) 26, 46, 125, 131, 145, 214
competition for land 9, 40, 119, 218
composting and food waste 180–181
concentrated animal feeding operations (CAFOs) 149–150, 174
conservation policies/programs: buffers 140; conservation promotion 167; exemptions 160; on farm conservation programs 171–173; irrigation organizations 179, *180*; soil health policies 169–171, *169–170*; water quality policies 173–176, *175*; water quality trading 176–178, *177*; water rights 178–179
Conservation Innovation Grants (CIG) 116
conservation land trusts 18
Conservation Reserve Enhancement Program (CREP) 115
Conservation Reserve Program (CRP) 115–116
Conservation Stewardship Program (CSP) 115–116
consolidation 7, 27–28; farm 6, 40, 97, 218; market 9

cooperative extension 60–61, 185–186, *185*, 222, 244
Corn Suitability Rating (CSR) 136
Coronavirus Aid, Relief, and Economic Security (CARES) Act 212
Covid-19 3, 25–26, 34, 67, 194–195, 211–212, 238
CSA *see* Community Supported Agriculture
CWA *see* Clean Water Act

Delaware 165–166
Department of Health and Human Services, the (HHS) 112, 206
Department of Homeland Security (DHS) 66
Department of Natural Resources (DNR) 222
Department of Transportation (DoT) 53, 106–107
Dillon's Rule state 53, 55
Direct-to-Consumer (DTC) 14, 19
Double Up Bucks program 203–204
Double Up Oklahoma (DUO) 204

economic development 19, 59, 67–68, 89, 98, 206; agricultural 63, 131; funds 160; staff 99
effective farm use (EFU) zoning 124
Electronic Benefit Transfer (EBT) 110, 193, 204
Emergency Food Assistance Program (TEFAP) 20, 110, 194, 212
emergency management/hazard mitigation plans 66
employment 13, 32, 67, 72, 200
Endangered Species Act of 1973 117
engagement 71, 80, 86–87, *88*, 128, 237
Environmental Protection Agency (EPA) 37, 117–119, 172, 174

INDEX

Environmental Quality Incentives Program (EQIP) 115–116
Environmental Working Group 37–38
EQIP *see* Environmental Quality Incentives Program
Erie County 64, *64*
Expanded Food and Nutrition Education Program, the (EFNEP) 206
Extension Disaster Education Network (EDEN) 66

Fair Food Network 203, 216n31
Farmable Wetlands Program 115
Farm Bill 20, 112, 114, 199; 1996 111; 2008 61; 2014 116, 199; 2018 199, 233; 2023 12; food programs 110; for nutrition assistance *108*; USDA and 107–108
Farmers Market Nutrition Program (FMNP) 110, 112, 193, 202
farmers markets 19, 112, 132, 145–147, 187; incubators and shared facilities 147
farming 166; farmland in 46; and food system 18, 206, 237; practices 12, 185; and ranching 119, 133, 158–159, 163
farm labor housing 142–143
farmland access 20, 119; access to farmland 218–219, *219*; buffers and setbacks 236; heirs property and fractionated lands 225–226; land access policy incentives 219–220; land banks 221–222; Land Buy Back Program for Tribal Nations 226; LandLink/FarmLink program 222; leasing public lands 222–223; Maryland's Next Gen program 220–221, *220*; mitigation fees 229–230; urban agriculture tax credits 223–225, *224*

farmland conversion 106, 119, 133, 158
farmland of statewide importance 44
farmland conservation: irrigation organizations 179, *180*; nutrient management plans 176; soil health policies 169–171, *169–170*; water quality policies 173–176, *175*; water quality trading 176–178, *177*; water rights 178–179
farmland protection policies/ programs 157–158; agricultural conservation easements 163; agricultural district programs 160–162, *161*; conservation easements 163; farmland mitigation 166–167; leasing development rights 160; property tax relief 158–159; purchase of agriculture conservation easement (PACE) 162–166, *162–163*, *165*; tax relief policies 159; use-value assessment 159; *see also* agricultural viability support and conservation policies/programs
Farmland Protection Policy Act (FPPA) 107
Farm Purchase and Protection programs 219–220
FarmLink programs 222
farms 17; city 71; family 25, 31–32; hydroponic and aquaponic 138; large 7, 37; midsized 6–7, 32, *33*, 218; planning for 97; rooftop 72; small commercial 7, 32, *33*, 138–139, 218; urban 152, 221, 223–225; U.S. 29
farms and food plan 25–27; cropland 32; disruptions in food 45–47; diverse family farms and rural economies 31–33, *33*; food insecurity and diet-related disease 33–35, *34*; huge gains

in productivity 27–31, 28; land use and land management 38–44, 39–43; natural resource degradation 35–37, 36

farms and food planning, principles and practices to: asset-based approach 79; asset mapping 91; carrying capacity and foodprint studies 91; community food assessments 91; developing and vetting the plan 95; Dot Poster Surveys 89; economic impact assessments 91; five Ws 94–95; foodshed analyses 91; food systems engagement process 93; food systems planning 80; future and set goals 92–94, 93; efforts guiding principles 78; goals and objectives 94–95; implement strategies and actions 95–96; in person or virtual **86**; planning practices 81; planning process 84–88, 88; plan with implementation 80; readiness 81–82, **82–84**; seven community capitals 79–80; "SMART" goals 94; trends and conditions 89–91, **90–91**

Farm Services Agency (FSA) 114–115, 167

farm viability 17, 114, 145, 163, 181, 186, 187

Fayette County 130–131

FDA *see* Food and Drug Administration

federal agencies 13, 20, 112, 116–117

Federal-Aid Highway Act 105–106

Federal Emergency Management Agency (FEMA) 66

federal government 9, 26, 53, 55, 106, 117, 205, 226

Federal Land Policy and Management Act of 1976 (FLPMA) 117

FFRI *see* Fresh Food Retailer Initiative

Fish and Wildlife Service 116

FLPMA *see* Federal Land Policy and Management Act

food access: Arizona's food policy council and 68; California, healthy-food access 69–70; in cities 13, 67–68, 238; Philadelphia's Food Trust 198; grant programs 110–111; for priority populations 194; in rural communities 4, 238

Food and Drug Administration (FDA) 112–113

Food Distribution Program on Indian Reservations 110, 206

food insecurity 10, 13, 31, 187, 192; and diet-related disease 33–35, 34; poverty and 146; rural 4

Food Quality Protection Act (FQPA) 117

Food Retail Expansion to Support Health (FRESH) program 143

food safety: conservation cost share programs 115–116; Conservation Reserve Program 115; easement programs 114; land retirement programs 115; partnership and grant programs 116–117

Food Safety and Inspection Service (FSIS) 112

Food Safety Modernization Act (FSMA) 113

food security 4, 9–10, 20, 71–72, 206; and cultural identity 12; definition 17; *see also* community food security

foodshed 18, 31, 70–71; analyses 91

food supply chain 18, 26, 61, 200

food systems: definition 18; emergency 211–212; issues 237–238; life cycle of food 8; planning 19, 67–69; planning for 70–74; plans 67–68; into sustainability planning 69–70; sustainable 18

food systems, agencies and policies 105–106; Bureau of Land Management 117; Environmental Protection Agency 117–119; federal role in land use policy 106, 107; food and nutrition 109–112; food safety 112–117; USDA and the Farm Bill 107–109, 108; see also food safety

food systems planning, overview of governmental levels 51–53; agriculture plans 63; climate action plans 65; common types of plans 61–74; comprehensive or master plans 62; cooperative extension 60–61; counties 55, 56, **57**; emergency management/hazard mitigation plans 66; food systems plans 67–68; Indigenous tribes 54–55; land use plans 61–62; local self-governance 55; metropolitan planning organizations 59; municipalities 57; planning bodies 59; planning commissions and boards 60; planning departments 60; planning for food systems at regional scale 70–74; regional planning commissions/councils of government 60; regional planning organizations 60; regional purpose authorities 60; regional transportation planning organizations 60; special districts 59; State Mitigation Planning Policy Guide 66; States 53–54; sustainability plans 69–70; towns and townships 58–59, **58**

food systems resiliency plans 68–69

Food Trust 198–199, *198*

fractionated lands 7, 218–219, 225–226

Fresh Food Retailer Initiative (FFRI) 200

Fresh Fruit and Vegetable Program 112

Fresno County, California 146

Fruit and Vegetable Prescription Program (FVRx) 205

FSMA *see* Food Safety Modernization Act

General Allotment (Dawes) Act 226

Geographic Information System (GIS) 81, 89

Good Agricultural Practices (GAP) 95

government structures 20

Grassland CRP 115

greenhouse gas (GHG) emissions 6, 11, *11*, 40–42, 45, 65, 218, 226

Gus Schumacher Nutrition Incentive Program, (GusNIP) 111, 193, 203–204

health 78; care 34, 40, 56; disparities 20, 195, 198; environmental 4, 13; human 5, 36–37, 43, 111; public 4, 13, 53, 112, 150, 183, 193; soil *see* soil health

Health and Human Services (HHS) 206

healthy corner store initiatives 196–197, *196*

healthy food financing initiatives (HFFI) 199–200

Healthy Food Healthy People working group 68, 75n24

healthy food retail 132, 143, 200

Healthy Incentives Program (HIP) 204

Healthy Savings 205

Heidel Hollow Farms 229

heirs property 7, 22n25, 31, 218–219, 225–226

HFFI *see* healthy food financing initiatives
historically Black colleges and universities (HBCUs) 61
"Home Rule" state 53, 55
housing 34, 39, 122; affordable 18, 44, 106, 137; farm labor 133, 142–143; seasonal 143
Hunger Free Oklahoma 204, 216n33
hypoxia 27–28, 36

Improving America's Schools Act 61
Indian Tribal Organizations (ITOs) 206
Intergovernmental Panel on Climate Change (IPCC) 10, 12, 26
International City/County Management Association (ICMA) 244

The Johns Hopkins Center for a Livable Future (CLF) 194, 244

Kaufman, Jerome 9
Keepseagle v. Vilsack 32

land banks 221–222
Land Buy Back program 226
LandLink program 222
Land Evaluation and Site Assessment (LESA) 153, 166, 230
landscapes 67, 78, 181, 230; agricultural 9, 40, *168*; American 106
land trust 18, 114, 163, 221, 230
land use planning 38, 47, 61–62, 105–106, 122, 124, 167, 194
legislation 137, 142, 169, 187, 194, 221, 223, 230; environmental 108; federal 66; historic 116
local and regional food systems (LRFS) 26, 46, 67, 193, 238, 245–246
local food 18–19, 98, 147, 186–187, 194; coop 99–100; organizations 98; production 65, 218, 221; sales 97
local governments 55; community-scale solar projects 227; mitigation fees 229–230; nutrition guidelines for public agencies 206–208; PACE programs 162–163; and states 53–54; on TEFAP 212
Louisiana Master Farmer Program 172
LRFS *see* local and regional food systems

Managerial and Professional Society (MAPS) 214
Marion County Public Health Department, Indiana 205
Maryland Agricultural and Resource-Based Industry Development Corporation (MARBIDCO) 184, 221
Maryland Department of Agriculture (MDA) 63, 74n8, 190n33
Massachusetts Department of Public Health 197, 215n17
Metropolitan Planning Organizations (MPOs) 59
mobile markets' 147, 194, 199, 201
mobile vending allowances 201–202
Morrill Act of 1890 61
Municipal Vulnerability Preparedness (MVP), Massachusetts 65, 75n15

National Agricultural Lands Study (NALS) 158
National Agricultural Law Center 244–245
National Agricultural Statistics Service (NASS) 17, 96
National Association of Conservation Districts (NACD) 245
National Association of Counties (NACo) 56, 245

National Association of Development Organizations (NADO) 245
National Association of Local Boards of Health, The (NALBOH) 245
National Association of Regional Councils (NARC) 245
National Association of Towns and Townships (NATaT) 245
National Conference of State Legislatures 54, 74n3, 240n40
National Institute of Food and Agriculture 81, 244
National Interstate and Defense Highways Act of 1956 106
National Oceanic and Atmospheric Administration's National Marine Fisheries Service 116
National Pollutant Discharge Elimination System (NPDES) 176
National School Lunch Program, the 206, 210
natural resource degradation 35–37, 36
natural resources 8, 60, 78–79, 117, 125, 163
Natural Resources Conservation Service (NRCS) 12, 44, 114–116, 136, 164
Navaho Nation 205, 208
New York State Energy Research and Development Authority (NYSERDA) 229
Nonpoint Source Management Program 118
nonpoint source pollution 118, 172, 175–176
North American Food System Network 245–246
NRCS see Natural Resources Conservation Service
nuisance complaints 119, 133, 181–182, 238

nutrition 9, 13, 37, 45, 108; guidelines 206–208, 207; incentives 196, 201–203; insecurity 26; security 13, 35, 205; see also nutrition education and promotion
nutritional security 13, 35
nutrition education and promotion 119, 194, 205–211
Nutrition Incentive Program (NIP) 111, 194, 202–205
nutritious food 17, 34, 84, 196, 213

Occupational Safety and Health Administration (OSHA) 66
Ohio River Basin Water Quality Trading Project 177
Option to Purchase at Agricultural Value (OPAV) 165
organic agriculture 17, 117, 125, 144, 224
"Our County" —Los Angeles Countywide Sustainability Plan 69–70

PACE see purchase of agricultural conservation easement
pesticides 5, 27, 36, 117
Pigford vs. Glickman 31
planning authority 20, 51, 53, 55, 77, 122, 125
planning organizations 20, 51
polycrisis 26–27, 238
Pothukuchi, Kami 9
Prairie Crossing Farm 125, 155n7
prime farmland 44, 136, 167
Produce Prescription Program (Veggie Rx) 111, 204
property rights 163–164; air rights 164; development rights 164; mineral rights 164; surface rights 164
public health 53, 60, 112, 150

Public Rangelands Improvement Act of 1978 117
purchase of agricultural conservation easement (PACE)/purchase of development rights (PDR) 126–127, 130, 162–166, *162–163*, *165*

REAP *see* Resource Enhancement and Protection
regenerative agriculture 9, 10, 17
Regional Conservation Partnership Program (RCPP) 116
regional planning organizations (RPOs) 60
regional purpose authorities (RPAs) 60
regional transportation planning organizations (RTPOs) 60
Reinvestment Fund 199
resilience 10, 26, 78, 218, 238; climate 72, 169; community 9; definition 19
Resource Enhancement and Protection (REAP) program 174
riparian buffers 140, 178
Roosevelt, Franklin D. 42, 50n59, 114
rural community 42; and agricultural viability 81; attributes of 72; food access in 238; impacts of farm consolidation 40; underserved 199
rural planning 13, 39, 73
Rural Urban Continuum (RUC) Codes 14

Safe Drinking Water Act 118
SALC *see* Sustainable Agricultural Lands Conservation
Saving Tomorrow's Agricultural Resources (STAR) program 170, 171
scenario planning 93
School Breakfast and Lunch program 112, 210

Schumacher, Gus 202–204, *203*
SCPEA *see* Standard City Planning Enabling Act
SCS *see* Soil Conservation Service
setbacks 140, 236
"SMART" goals 94
smart growth 9, 39, 47, 118, 168, 237
Smart Solar^{SM} 232–233, *232*; solar on disturbed and developed sites 228–229, *229*
Smith-Lever Act 60, 185
SNAP *see* Supplemental Nutrition Assistance Program
Soil Conservation Service (SCS) 114
soil health: and ecosystems 17; and environmental management 237; investments in 223; policies 169–171, *170*; and productivity 228
soil suitability 45
Soil Survey Geographic Database (SSURGO) 44, **91**
solar energy 226–228, *227*
Sonoma County, California 182, 190n43
Special Supplemental Nutrition Program for Women, Infants, and Children (WIC) 112, 206
Standard City Planning Enabling Act (SCPEA) 61–62
Standard State Zoning Enabling Act (SZEA) 61, 131
State Acres for Wildlife Enhancement (SAFE) 115
subdivision regulations 124–126; Agrihoods 125; cluster development 126
Supplemental Nutrition Assistance Program (SNAP) 4, 34, 54, 109–111, *110*, 192–194, 196–198
Sustainable Agricultural Lands Conservation (SALC) Program 168–170
sustainable agriculture 17, 20, 46, 195

Sustainable Development Code 246
sustainable food systems 18, 78, 87, 121
sustainability 19; plans 69–70
Sweetened Beverage Tax 208, 212
SWOT analysis 89, 95

Tampa Bay Regional Resilience Action Plan 65
TEFAP *see* Emergency Food Assistance Program
The UConn Rudd Center for Food Policy and Health 215n5, 246
The Sustainable Development Code 246
Total Maximum Daily Load (TMDL) 174
towns and townships governments 58–59, **58**
transfer of development rights (TDR) programs (Transfer of Development Credits and Transferable Development Units) 126–128, *127*, 137, 160
transformation 12, 16, 20, 47, 237
Transition Incentives Program (TIP) 115
transportation 11, 53, 61, 122; rural 60
Transportation Management Areas (TMAs) 59

Uniform Partition of Heirs Property Act (UPHPA) 225
unique farmland 44
United States Conference of Mayors 246
UPHPA *see* Uniform Partition of Heirs Property Act
urban agriculture 9, 71–72, *71*, 194, 214
urban agriculture zones 137–139, *138*
urban growth boundaries (UGBs)/ urban service boundaries 128–131, *129*

USDA Food and Nutrition Service 21n7, 49n38, 120n9, 193, 215n14, 215n18
USDA's Forest Service 106, 109
U.S. Composting Council 142, 156n29
U.S. Department of Energy (DOE) 41, 50n57, 218, 227, 239n1, 240n25
U.S. Fish and Wildlife Service 116

Vermont Farm to Plate (F2P) 67, 96
Vermont Law School Center for Agriculture and Food Systems 119n2, 246
Vermont's Clean Water Act (Act 64) 175
Vermont Sustainable Jobs Fund (VSJF) 75
visioning 93

water quality policies 173–176, *175*
water quality trading 176–178, *177*
water rights 178–179
Wayne County, farms and food in 97–100, *97*
wetland reserve easements (WRE) 114
Wetland Reserve Enhancement Partnership 114
World Café 85–87

zoning: accessory and ancillary uses of agricultural land 139; agricultural 132–133; agricultural preservation 135–136; agricultural protection zones 133–134; for agriculture in Scott County 135–136, *135*; agritourism 139–140; composting 141–142, *142*; concentrated animal feeding operations 149; considerations 139; effective/ exclusive agricultural 134; Euclidean zoning 131–132; farmers' markets 145–147, *146*; farm labor housing 142–143; form-based zoning 132; healthy food retail 143; incentive zoning 132; incubators and shared facilities 147; intensity zoning 132; intensive agricultural 134; local food 143–144; mixed-use zoning 132; on-farm energy production 148–149, *148*; on-farm marketing 144–145, *144*; ordinances 132; overlay 137; performance zoning 132; poultry and livestock 149–150; setbacks and buffers 140–141; signage 152; sliding-scale 134; and subdivision regulations 124–125; urban agriculture 137–139, *138*; urban chickens, bees, and livestock 151–152, *151*; wineries, breweries, and distillers 152–154, *153*; zoning ordinances 132

For Product Safety Concerns and Information please contact our EU
representative GPSR@taylorandfrancis.com
Taylor & Francis Verlag GmbH, Kaufingerstraße 24, 80331 München, Germany